The
MASTER HANDBOOK
Of
IC CIRCUITS
2nd Edition

No. 3185
$34.95

The
MASTER HANDBOOK
Of
IC CIRCUITS
2nd Edition

Delton T. Horn

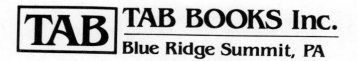

TAB BOOKS Inc.
Blue Ridge Summit, PA

SECOND EDITION
FIRST PRINTING

Copyright © 1989 by TAB BOOKS Inc.
Printed in the United States of America

Library of Congress Cataloging in Publication Data

Horn, Delton T.
The master handbook of IC circuits / by Delton T. Horn. — 2nd ed.
p.   cm.
Rev. ed. of: The master handbook of IC circuits / by Thomas R.
Powers. 1st ed. c1982.
Includes index.
ISBN 0-8306-9185-5   ISBN 0-8306-3185-2 (pbk.)
1. Integrated circuits—Handbooks, manuals, etc.   I. Powers,
Thomas R.   Master handbook of IC circuits.   II. Title.
TK7874.H678   1989
621.381'73—dc19   88-32976
                                                          CIP

TAB BOOKS Inc. offers software for
sale. For information and a catalog,
please contact TAB Software Department,
Blue Ridge Summit, PA 17294-0850.

Questions regarding the content of this book
should be addressed to:

Reader Inquiry Branch
TAB BOOKS Inc.
Blue Ridge Summit, PA 17294-0214

Edited by Eric Phelps

Contents

1.2 to 37 Volt Regulator • 337T 1.2 to 37 Volt Regulator • 340 Series Voltage Regulator • 342 Integrated Voltage Regulator • TL431 Adjustable Shunt (Zener) Regulator • 0070 Precision BCD Buffered Voltage Reference • 7805 Five Volt Integrated Voltage Regulator • 78L05 Five Volt Integrated Voltage Regulator • 7812 Twelve Volt Integrated Voltage Regulator • 7815 Fifteen Volt Integrated Voltage Regulator • 7905 Negative Five Volt Series Voltage Regulator

Introduction

No development in electronics has ever had the impact of the integrated circuit. That's a strong statement, considering such prior developments as the vacuum tube and transistor. But you'll probably be convinced of its truth as you scan through the pages of this book!

You're already familiar with such IC advantages as small size and low power requirements. What you may be unaware of is how ICs greatly simplify circuit design. Many ICs have been designed with specific purposes in mind, in contrast to many discrete components. We've gathered together over 900 applications circuits using over 200 popular ICs. The result is a smorgasbord of ideas and designs. All you have to do is connect the appropriate components to each IC and you're in business.

We've used a generic numbering system to refer to the ICs in this book. Various manufacturers of devices will add their company prefix (such as LM for National, SN for Texas Instruments, CA for RCA, etc.) to some of the devices in this book. Operationally, a device identified by a certain number will be identical from manufacturer to manufacturer.

Some of the circuits in this book may have a resistor and/or capacitor or two with no value indicated. The values of such parts should be determined experimentally for best circuit operation. Other circuits may have unlabeled transistors. In such circuits, the transistor can be any general purpose PNP or NPN transistor, as indicated.

With many ICs now available on the surplus market at amazingly low prices, you may have wanted to use more of them in your electronic projects and experiments but been stymied by the lack of application information. That's the void this book hopes to fill. So go scrounging around in your junk box—odds are you'll find a couple of ICs you thought were useless, but there are some circuits using them in this book!

In this second edition, many new ICs and circuits have been added and obsolete devices have been eliminated. The new edition also includes many pin-out diagrams for popular ICs and some explanatory text, where appropriate. The text has been kept to an absolute minimum. The purpose of this book is to provide the maximum number of practical circuits with a minimum of fuss and bother.

Part 1
OP AMPS

Undoubtedly the most popular and widely used type of IC is the op amp. These range from inexpensive, general purpose devices to specialized high-grade units. This section offers dozens of applications for this class of IC.

101—GENERAL PURPOSE OPERATIONAL AMPLIFIER

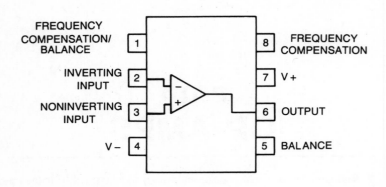

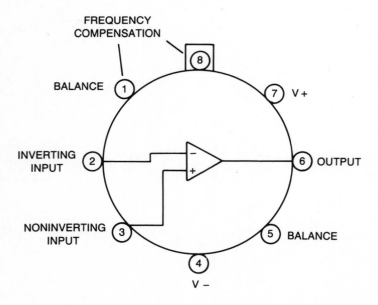

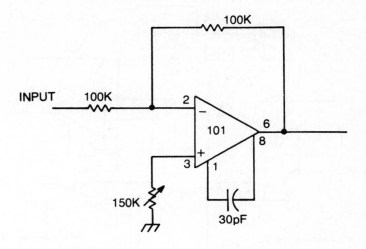

Summing amplifier with bias-current compensation

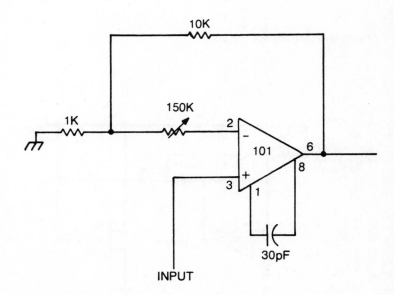

Noninverting amplifier with bias-current compensation

3

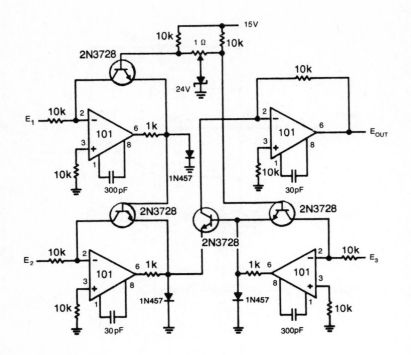

Analog multiplier/divider

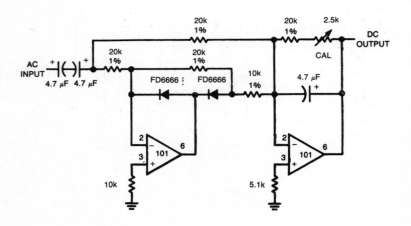

Full-wave rectifier and averaging filter

4

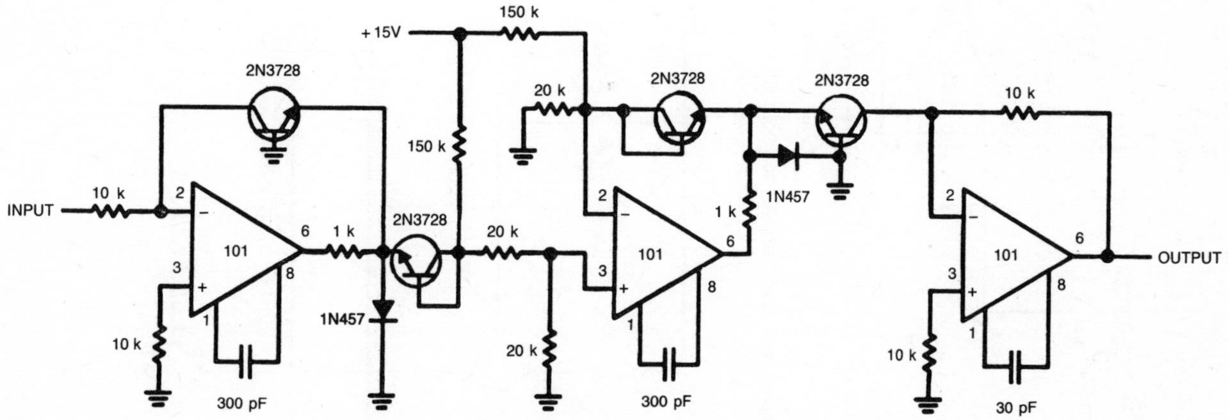

Root extractor

5

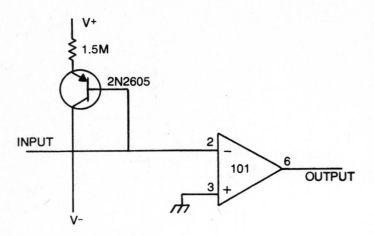

Summing amplifier with bias-current compensation and improved temperature stability

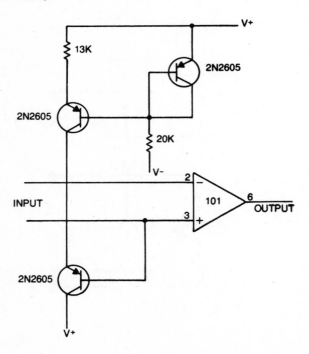

Bias-current compensated noninverting amplifier operating over large common mode range

6

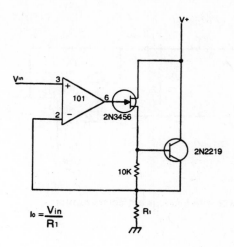

$$I_0 = \frac{V_{in}}{R_1}$$

Precision current sink

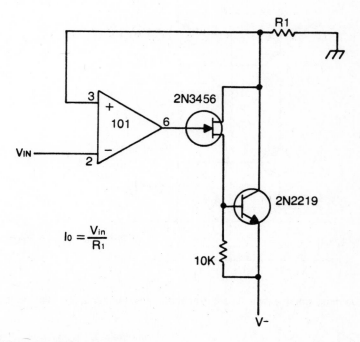

$$I_0 = \frac{V_{in}}{R_1}$$

Precision current source

7

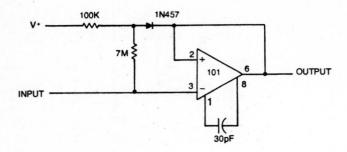

Voltage follower with bias-current compensation

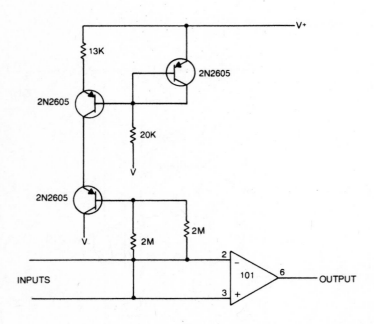

Summing amplifier with bias-current compensation for differential inputs

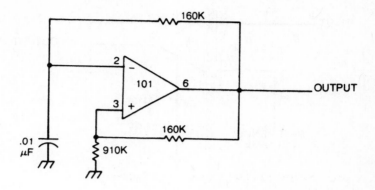

Low frequency free-running multivibrator

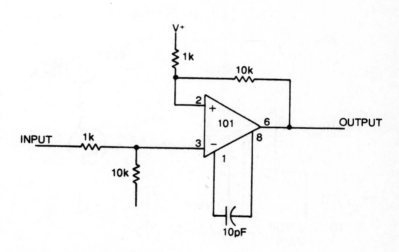

Level-shifting differential amplifier

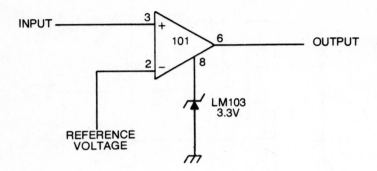

Voltage comparator for driving TTL ICs

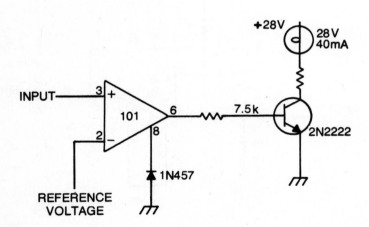

Voltage comparator and lamp driver

10

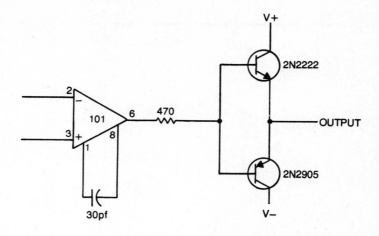

High current output buffer

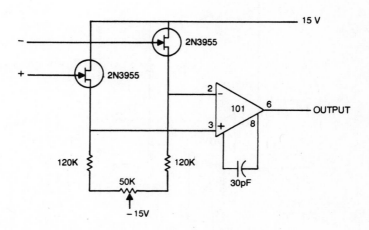

Summing amplifier with FET source followers

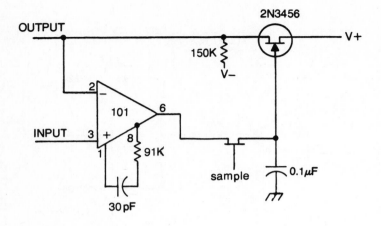

Low drift sample and hold

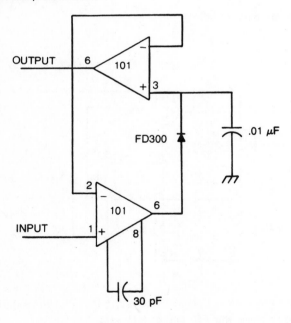

Positive peak detector with buffered output

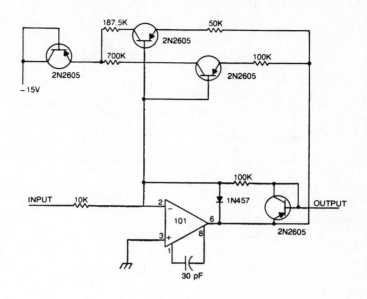

Nonlinear amplifier with temperature compensated breakpoints

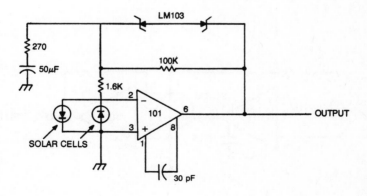

Saturating servo control preamplifier with rate feedback and solar cell sensors

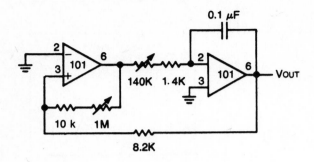

Triangular-wave generator

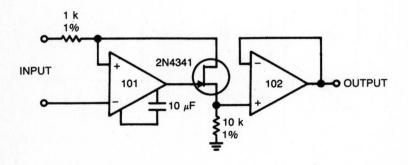

Level-shifting isolation amplifier

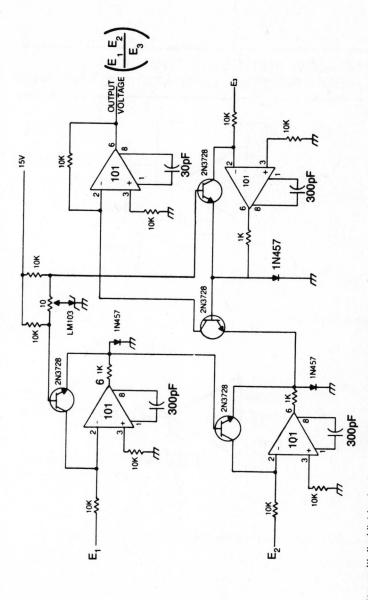

Analog multiplier/divider for input voltages from 500 mV to 50 V

15

101A—LOW INPUT CURRENT
GENERAL PURPOSE OPERATIONAL AMPLIFIER

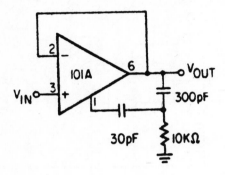

Fast voltage follower

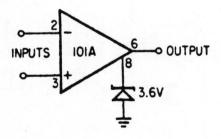

Voltage comparator for driving DTL or TTL integrated circuits

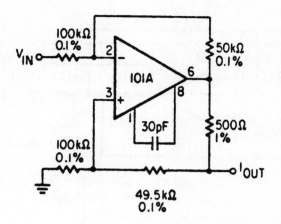

Bilateral current source

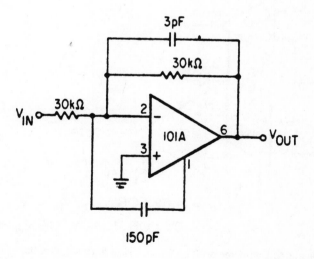

Fast summing amplifier

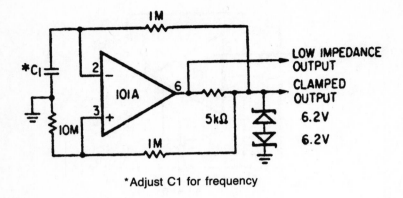

*Adjust C1 for frequency

Low frequency square-wave generator

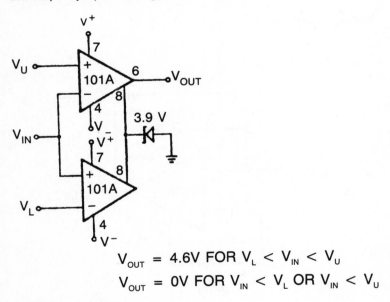

$$V_{OUT} = 4.6V \text{ FOR } V_L < V_{IN} < V_U$$
$$V_{OUT} = 0V \text{ FOR } V_{IN} < V_L \text{ OR } V_{IN} < V_U$$

Double-ended limit detector

18

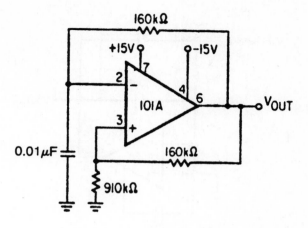

Free-running multivibrator

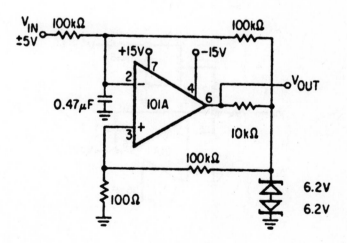

Pulse width modulator

19

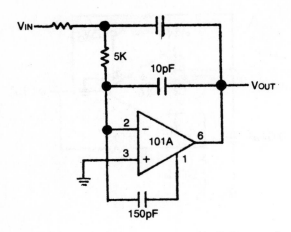

Fast integrator

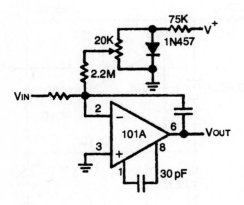

Integrator with bias current compensation

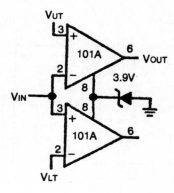

Double-ended limit detector

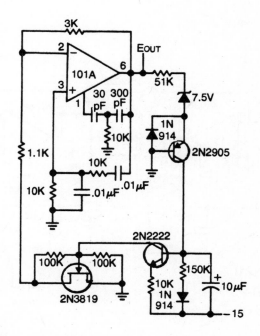

Wien bridge oscillator with FET amplitude stabilization

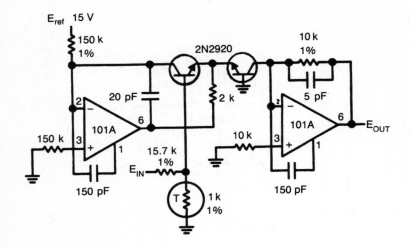

Anti-log generator

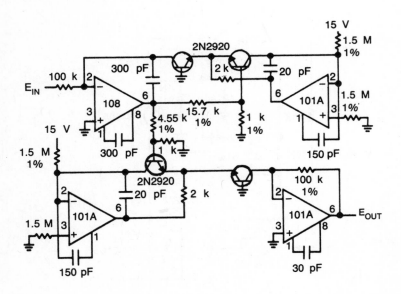

Cube generator

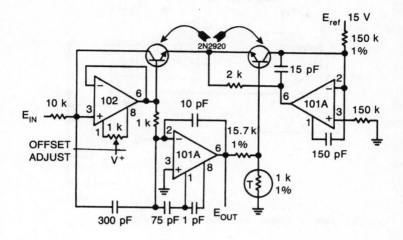

Fast log generator

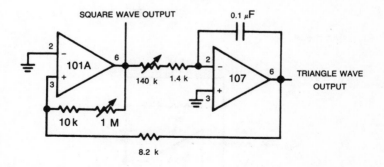

Function generator

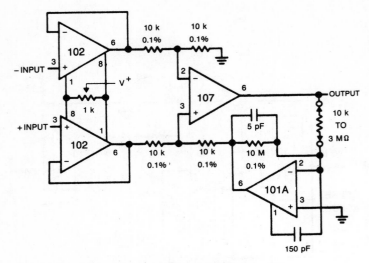

Variable gain, differential-input instrumentation amplifier

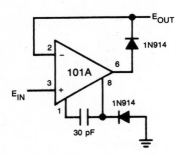

Precision diode

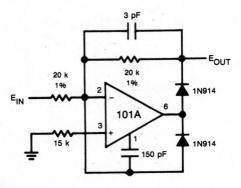

Fast half-wave rectifier

24

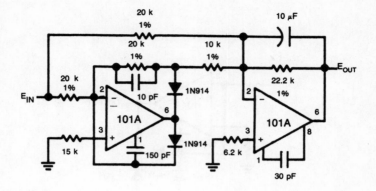

Precision ac to dc converter

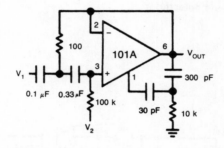

Tuned circuit

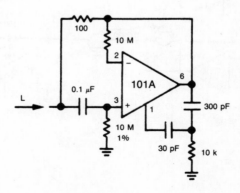

Simulated inductor

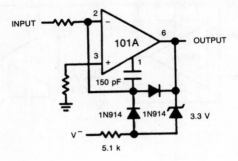

Fast zero crossing detector

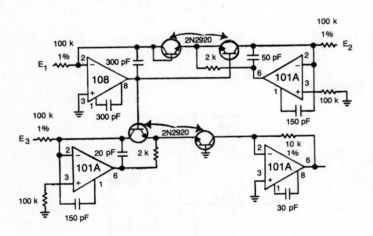

Multiplier/divider

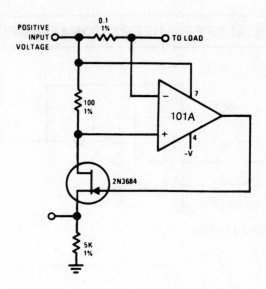

Current monitor

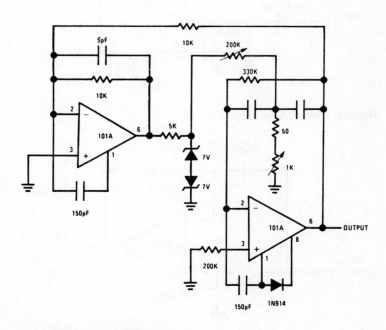

Low distortion sine-wave oscillator

102—HIGH SPEED OPERATIONAL AMPLIFIER

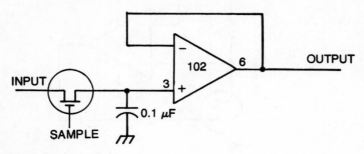

Sample and hold circuit

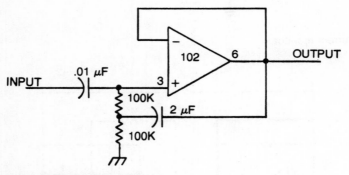

High input impedance ac amplifier

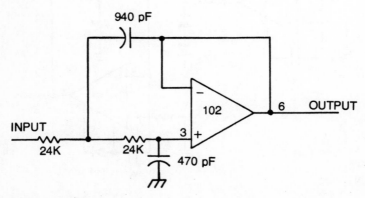

Low-pass active filter

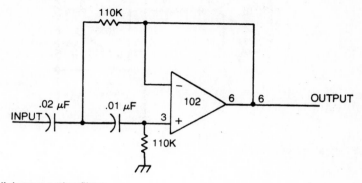

High-pass active filter

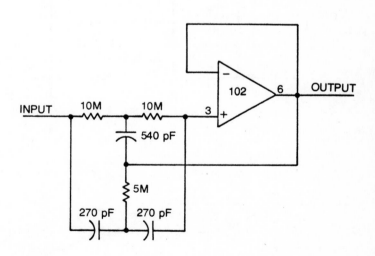

High Q notch filter

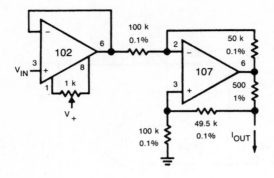

Bilateral current source

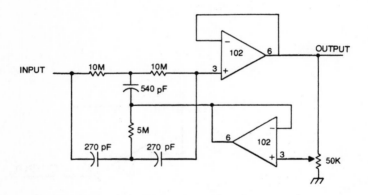

Variable Q notch filter

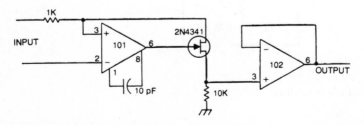

Level-shifting isolation amplifier

30

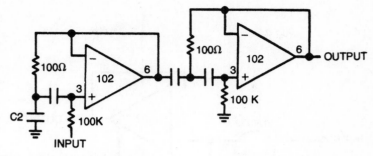

Two-stage tuned circuit

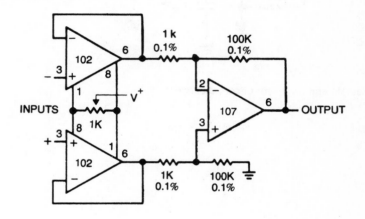

Differential-input instrumentation amplifier

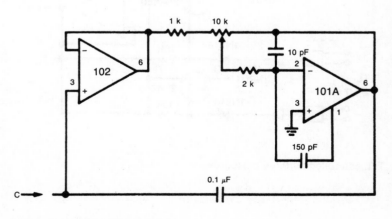

Variable capacitance multiplier

31

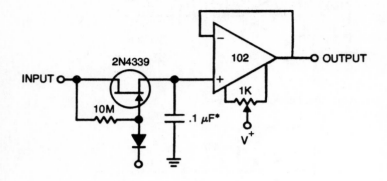

Sample and hold with offset adjustment

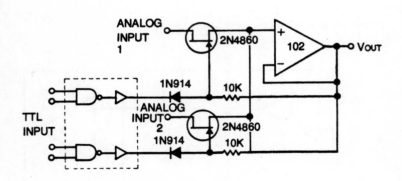

TTL controlled buffered analog switch

124—QUAD OPERATIONAL AMPLIFIER

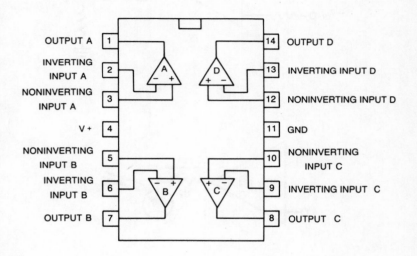

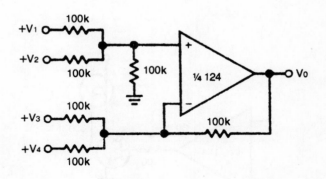

DC summing amplifier

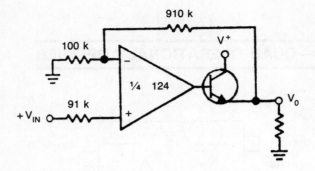

Power amplifier

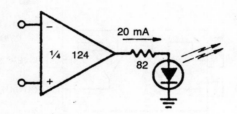

LED driver

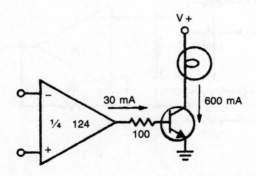

Lamp driver

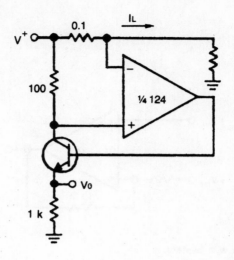

Current monitor

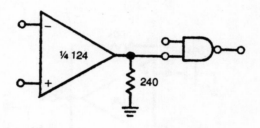

Driving TTL

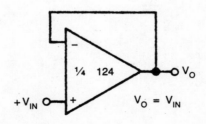

Voltage follower

35

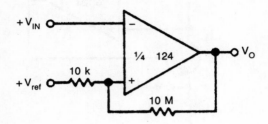

Comparator with hysteresis

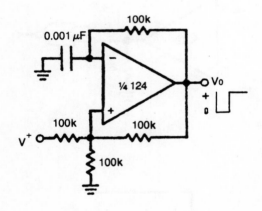

Square-wave oscillator

36

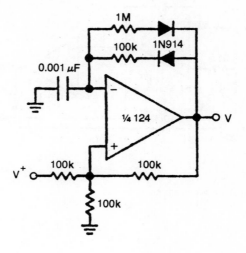

Pulse generator

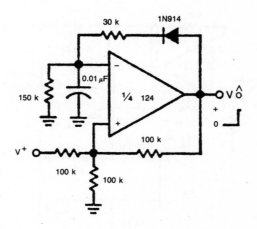

Pulse generator

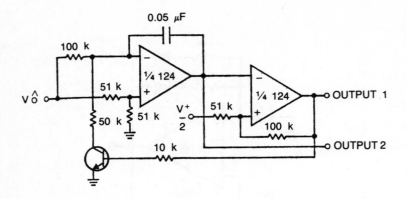

Voltage controlled oscillator circuit

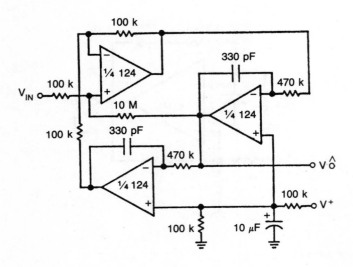

"BI-QUAD" RC active bandpass filter

38

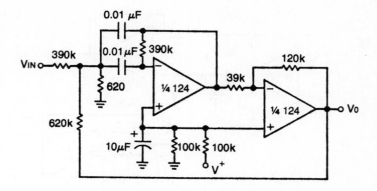

Bandpass active filter

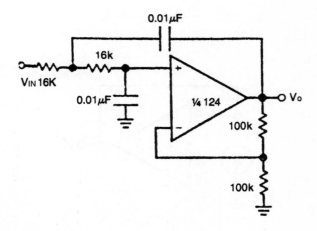

DC coupled low-pass RC active filter

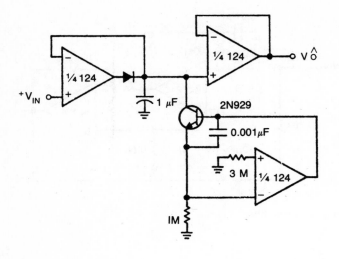

Low drift peak detector

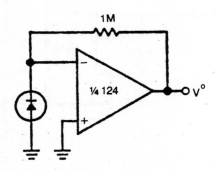

Photo voltaic-cell amplifier

40

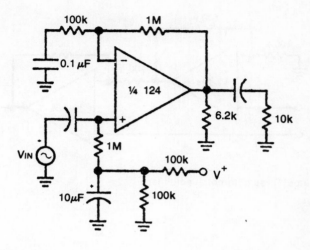

AC coupled noninverting amplifier

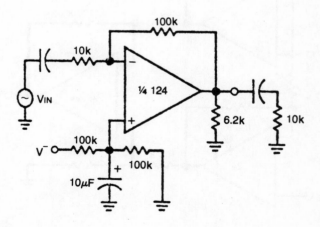

AC coupled inverting amplifier

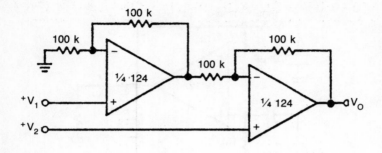

High input Z, dc differential amplifier

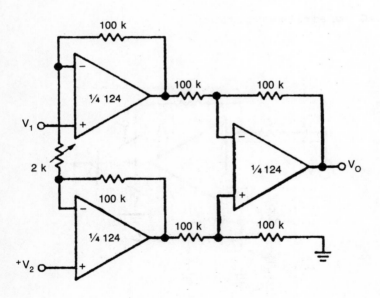

High input Z adjustable-gain dc instrumentation amplifier

143—HIGH VOLTAGE OPERATIONAL AMPLIFIER

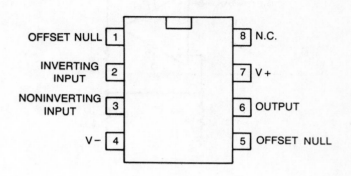

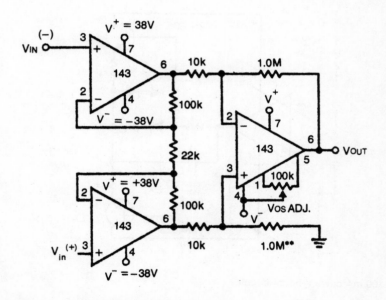

Common-mode instrumentation amplifier

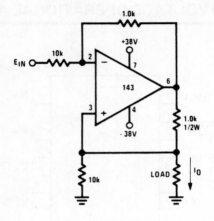

High-compliance current source

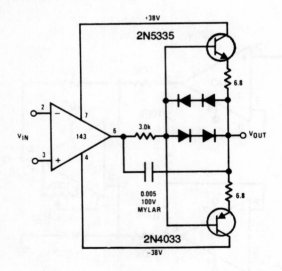

100 mA current boost circuit

All diodes are 1N914

144—HIGH VOLTAGE, HIGH SLEW RATE OPERATIONAL AMPLIFIER

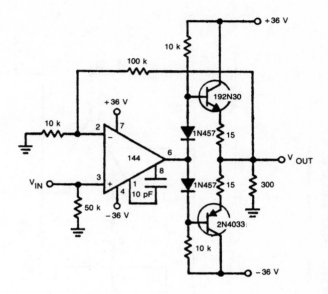

Large power bandwidth, current boosted audio line driver

146—PROGRAMMABLE QUAD OPERATIONAL AMPLIFIER

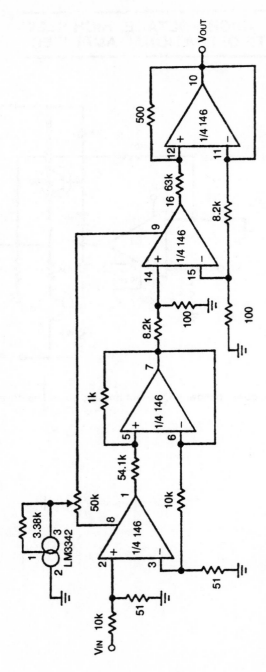

A 4th order Butterworth low-pass capacitorless filter

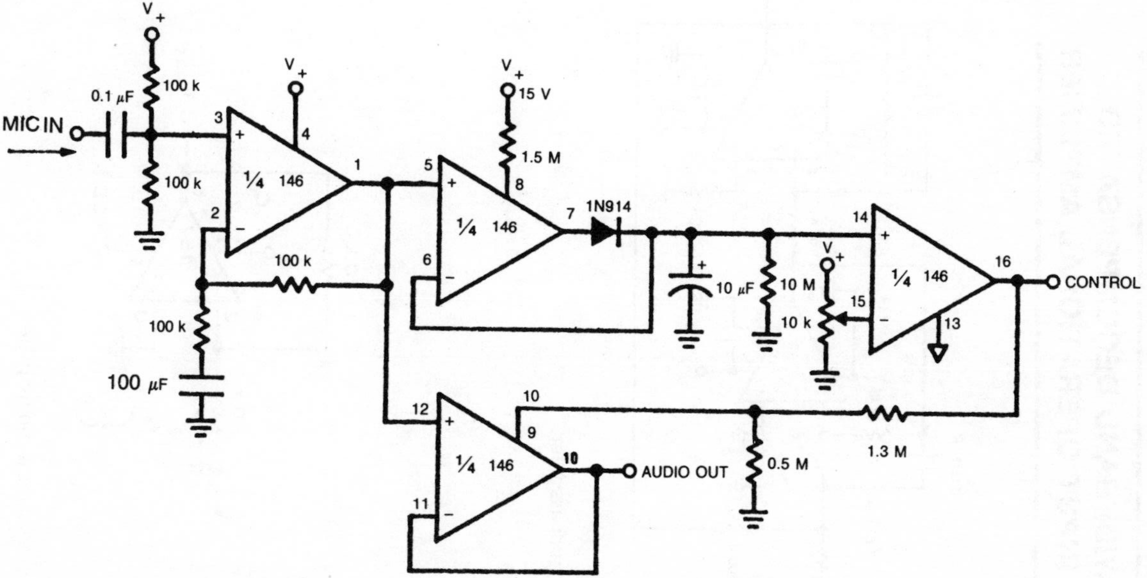

Voice activated switch and amplifier

47

157—WIDEBAND DECOMPENSATED JFET INPUT OPERATIONAL AMPLIFIER

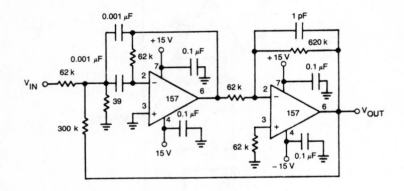

High Q bandpass filter

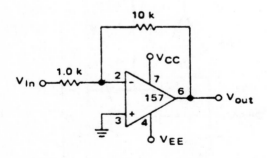

Large power bandwidth amplifier

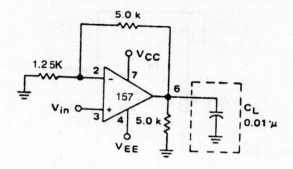

Driving capacitive loads

158—LOW POWER DUAL
OPERATIONAL AMPLIFIER

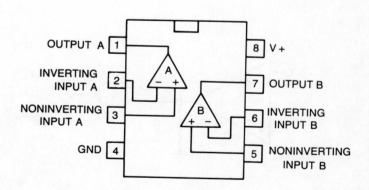

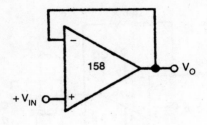

Voltage follower

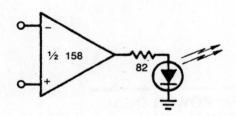

LED driver

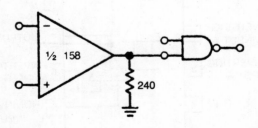

Driving TTL

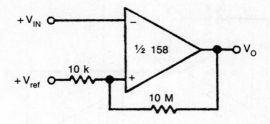

Comparator with hysteresis

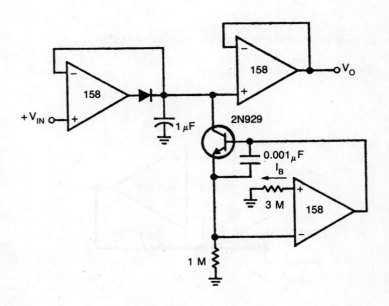

Low drift peak detector

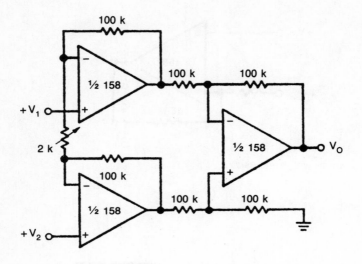

High input Z adjustable-gain dc instrumentation amplifier

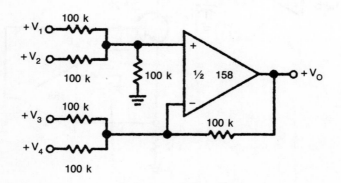

DC summing amplifier

52

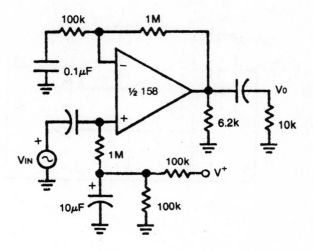

AC coupled noninverting amplifier

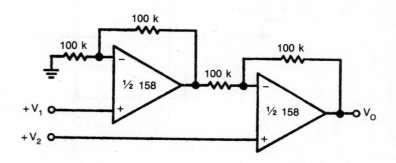

High input Z, dc differential amplifier

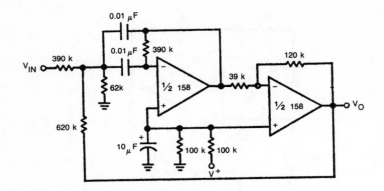

Bandpass active filter

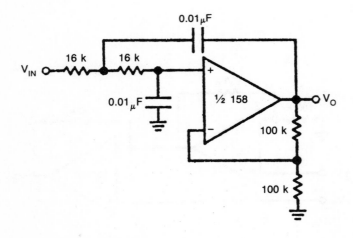

DC coupled low-pass RC active filter

54

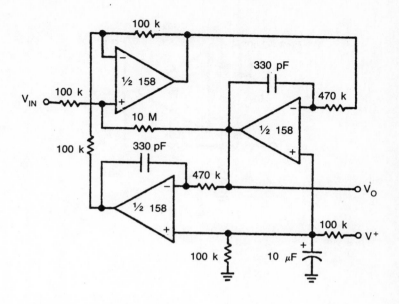

"BI-QUAD" RC active bandpass filter

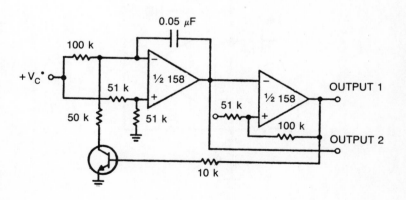

Voltage controlled oscillator (VCO)

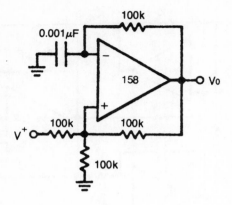

Square-wave oscillator

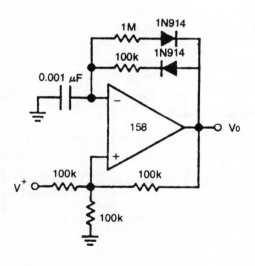

Pulse generator

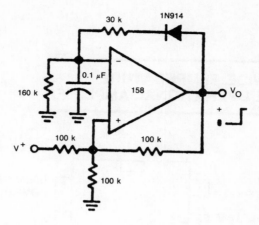

Pulse generator

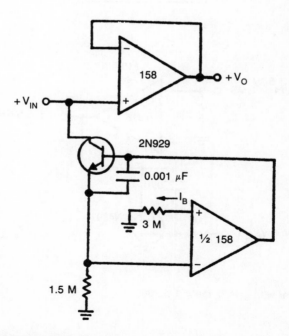

Using symmetrical amplifiers to reduce input current (general concept)

57

318—WIDE TEMPERATURE RANGE OPERATIONAL AMPLIFIER

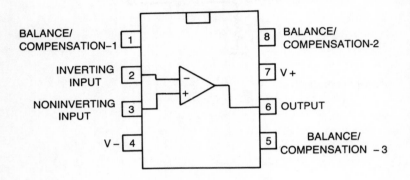

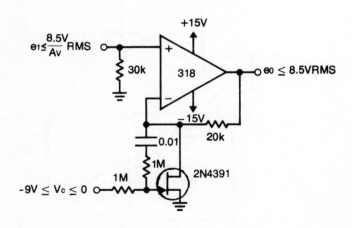

Amplifier with gain range = 1 – 1000

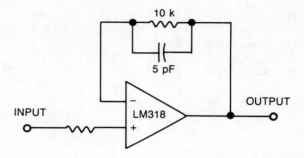

Fast voltage follower

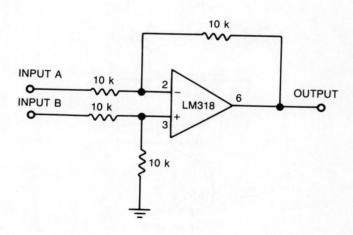

59

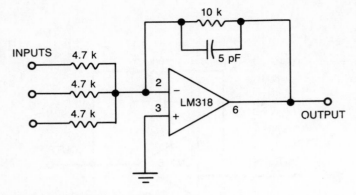

Fast inverting summing amplifier

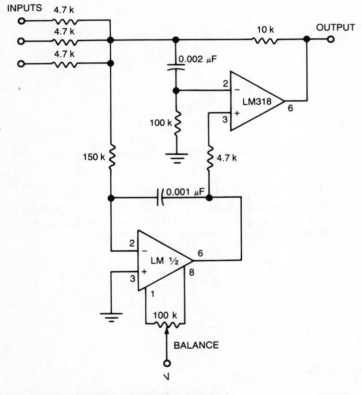

Fast summing amplifier with low input current

60

324—QUAD OP AMP

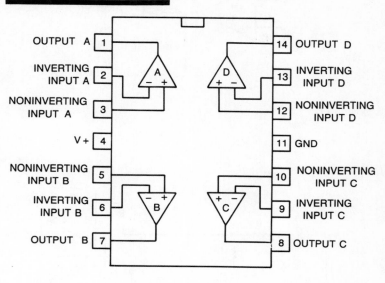

OUTPUT A — 1	14 — OUTPUT D
INVERTING INPUT A — 2	13 — INVERTING INPUT D
NONINVERTING INPUT A — 3	12 — NONINVERTING INPUT D
V+ — 4	11 — GND
NONINVERTING INPUT B — 5	10 — NONINVERTING INPUT C
INVERTING INPUT B — 6	9 — INVERTING INPUT C
OUTPUT B — 7	8 — OUTPUT C

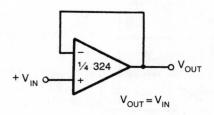

$+V_{IN}$ ¼ 324 V_{OUT}

$V_{OUT} = V_{IN}$

Voltage follower

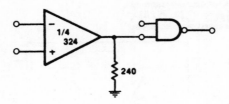

Driving TTL

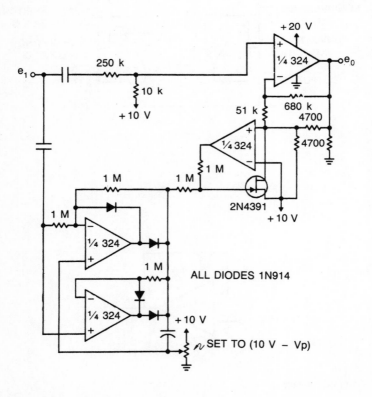

Volume expander circuit

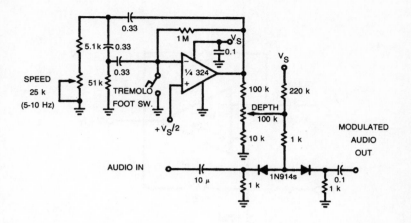

Tremolo circuit

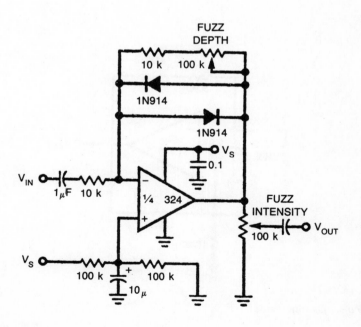

"Fuzz" circuit

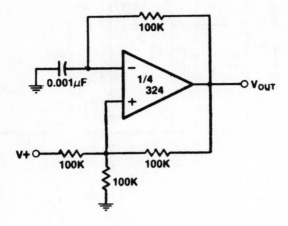

Square-wave oscillator

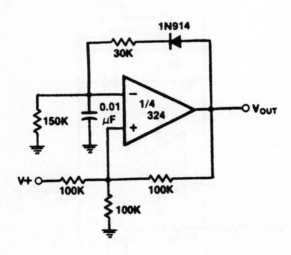

Pulse generator

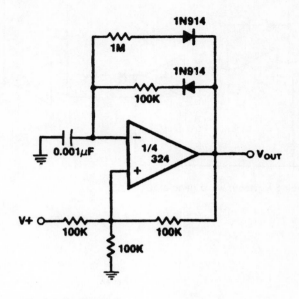

Pulse generator

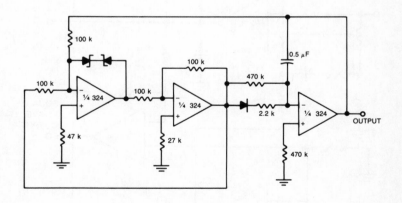

Ascending sawtooth wave generator

65

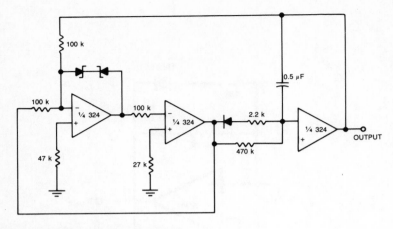

Descending sawtooth wave generator

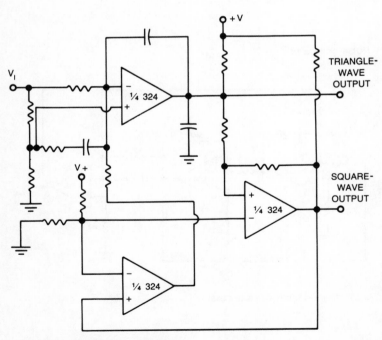

VCO (voltage-controlled oscillator)

66

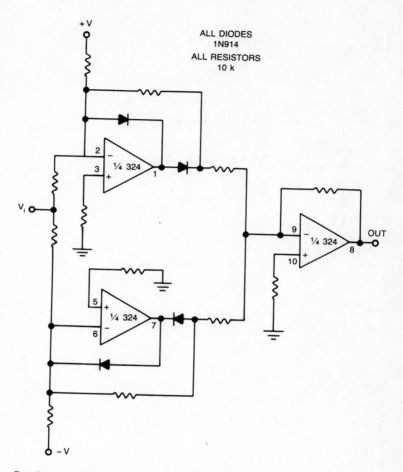

+V

ALL DIODES
1N914
ALL RESISTORS
10 k

V$_i$

2
3
¼ 324
1

5
+
6
¼ 324
7

9
10
+
¼ 324
8

OUT

−V

Dead space circuit

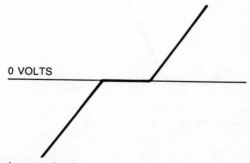

0 VOLTS

Output of dead space circuit

67

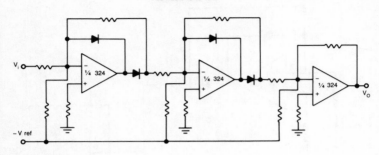

Series limiter

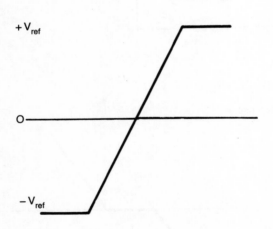

Output of series limiter

347—WIDE BANDWIDTH QUAD JFET INPUT OPERATIONAL AMPLIFIER

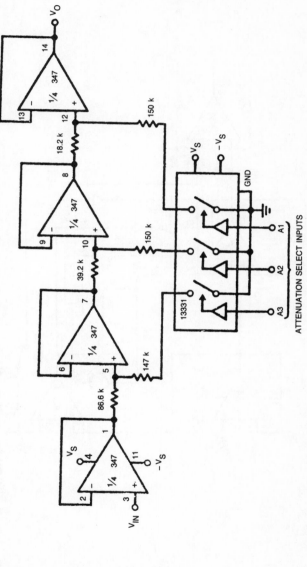

Digitally selectable precision attenuator

69

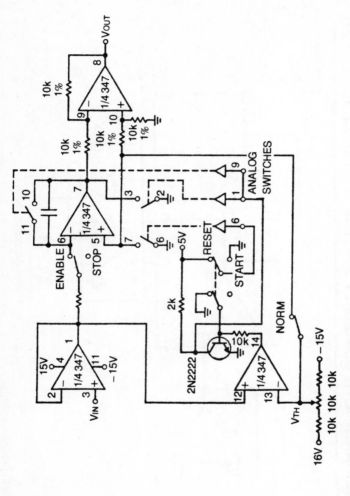

Long time integrator with reset, hold and starting threshold adjustment

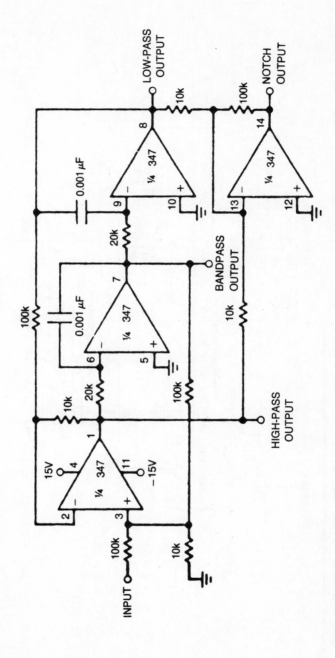

Universal state variable filter

71

348—QUAD 741 OPERATIONAL AMPLIFIER

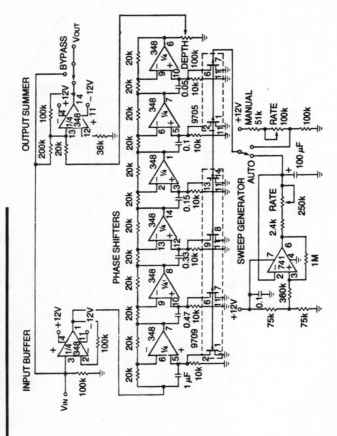

Phase Shifter

349—WIDEBAND DECOMPENSATED QUAD 741 OPERATIONAL AMPLIFIER

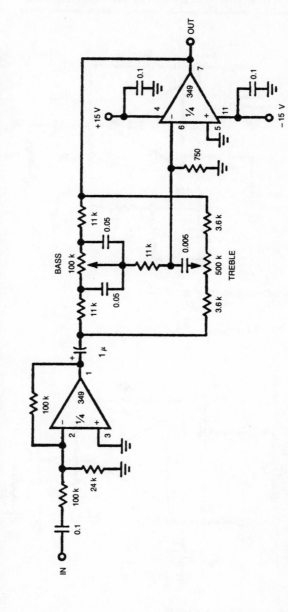

Active bass & treble tone control with buffer

353—WIDE BANDWIDTH DUAL JFET INPUT OPERATIONAL AMPLIFIER

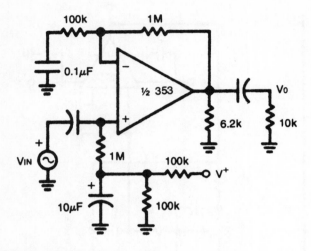

AC coupled noninverting amplifier

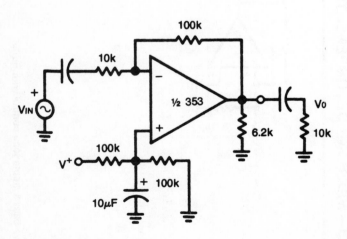

AC coupled inverting amplifier

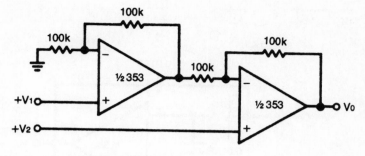

High input Z, dc differential amplifier

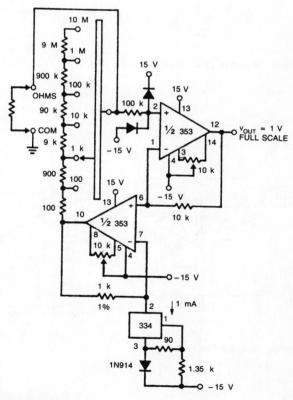

Ohms to volts converter

75

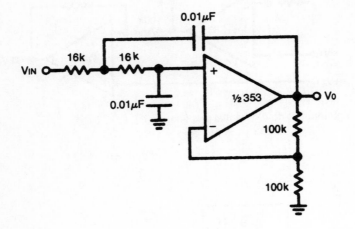

DC coupled low-pass RC active filter

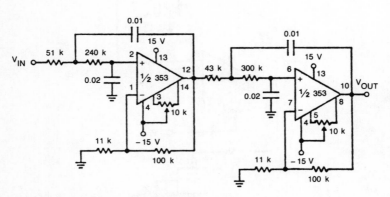

Fourth order low-pass Butterworth filter

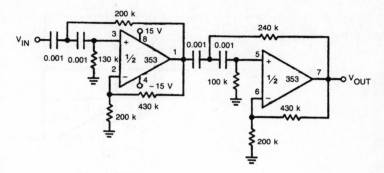

Fourth order high-pass Butterworth filter

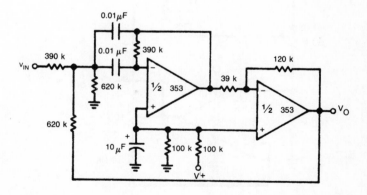

Bandpass active filter

77

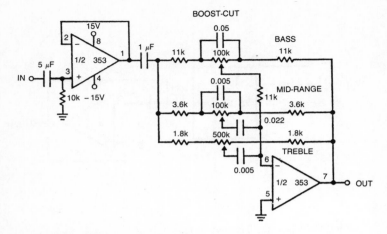

Three-band active tone control

356—LOW OFFSET MONOLITHIC
JFET INPUT OPERATIONAL AMPLIFIER

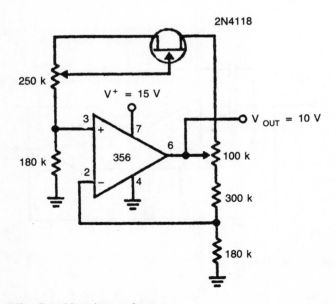

Low drift adjustable voltage reference

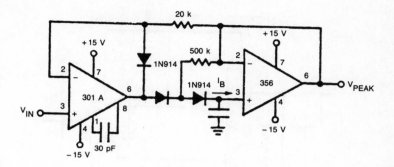

Low drift peak detector

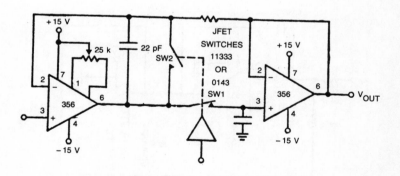

High accuracy sample and hold

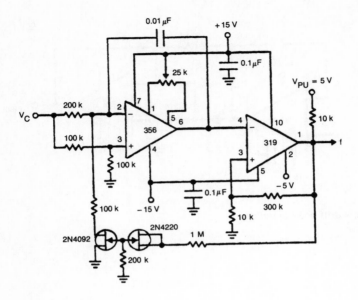

3 decade

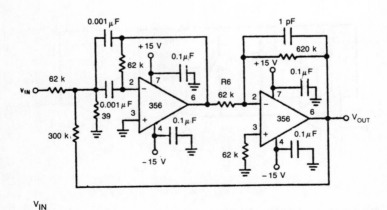

v_{IN}

High Q bandpass filter

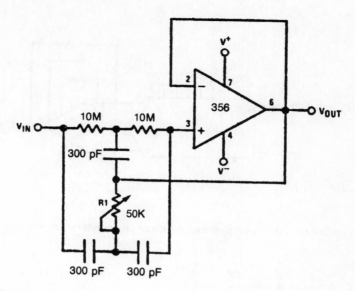

High Q notch filter

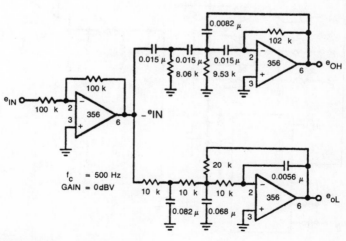

Active crossover network

81

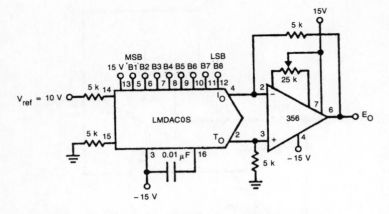

8-bit D/A converter with symmetrical offset binary operation

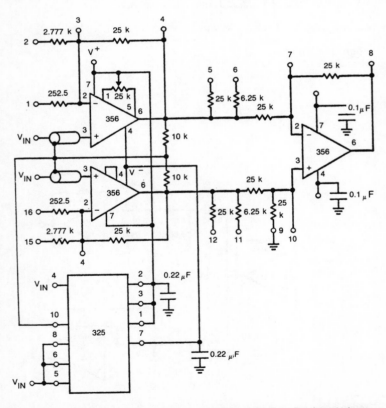

High CMRR, low drift instrumentation amplifier with floating input stage

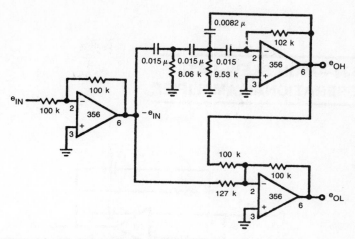

Asymmetrical 3rd order Butterworth active crossover network

357—MONOLITHIC JFET INPUT OPERATIONAL AMPLIFIER

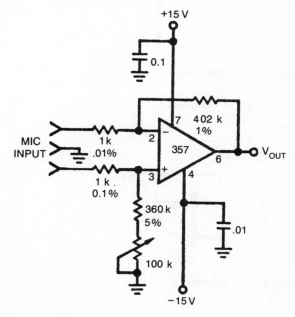

Transformerless mic preamp for balanced inputs

531—HIGH SLEW RATE OPERATIONAL AMPLIFIER

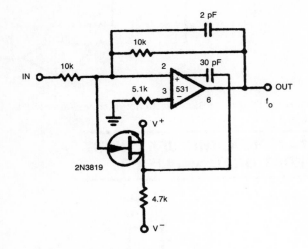

High speed inverter (10MHz bandwidth)

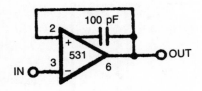

Fast settling voltage follower

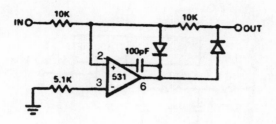

Half-wave rectifier

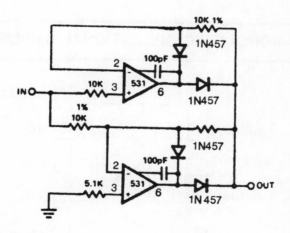

Full-wave rectifier

85

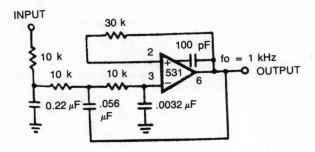

3 pole active low-pass filter

709—MONOLITHIC OPERATIONAL AMPLIFIER

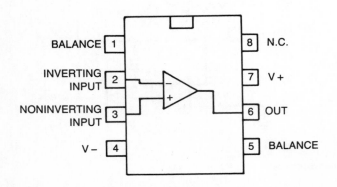

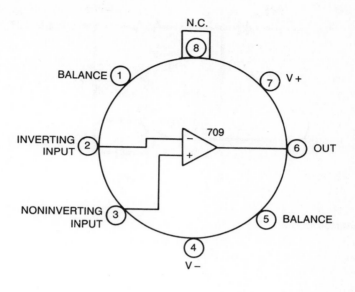

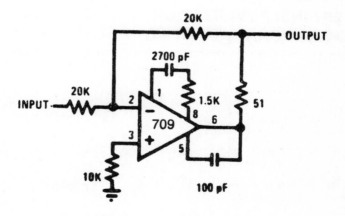

Unity gain inverting amplifier

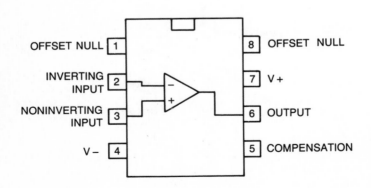

Voltage follower

725—INSTRUMENTATION OPERATIONAL AMPLIFIER

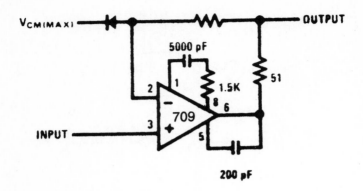

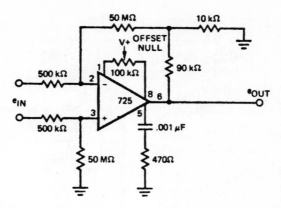

Precision amplifier

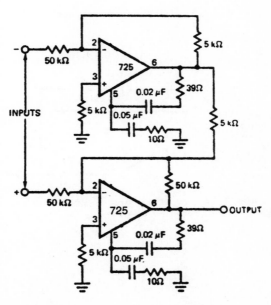

$\pm$ 100 V common mode range differential amplifier

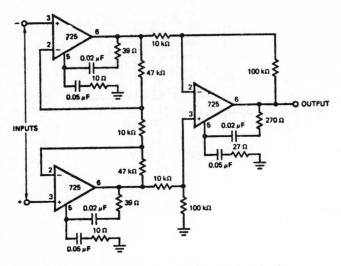

Instrumentation amplifier with high common mode rejection

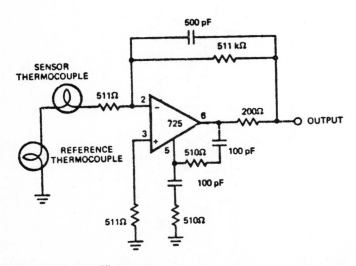

Thermocouple amplifier

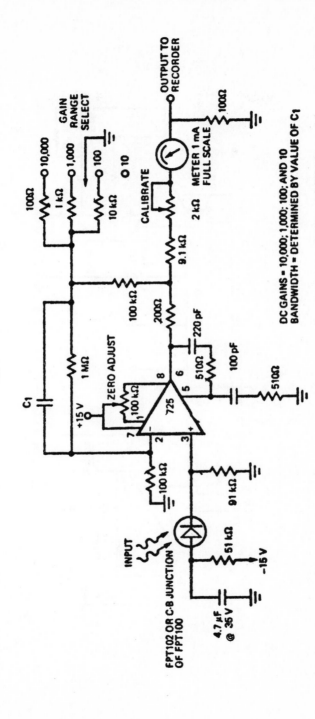

Photodiode amplifier

91

739—DUAL LOW NOISE AUDIO AMPLIFIER/OPERATIONAL AMPLIFIER

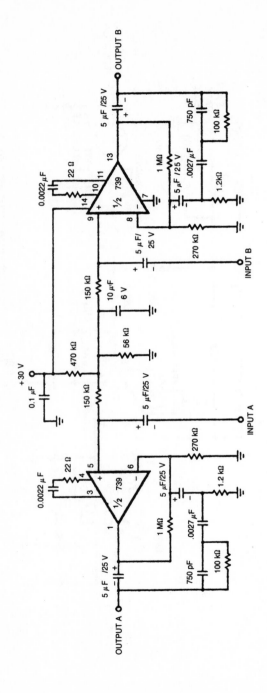

Stereo phono preamplifier — RIAA equalized

741—FREQUENCY COMPENSATED OPERATIONAL AMPLIFIER

The 741 is probably the best selling IC of all time. It can be used in countless applications. It has become the de facto standard for all op amp ICs. Other op amp chips are compared to the 741. Most op amp chips are pin-for-pin compatible with the 741.

When the 741 first came out, it was considered a high-grade device. Since then, the technology has improved considerably, and the 741 is now used primarily in noncritical and experimental applications. It is generally too noisy for serious audio or precision measurement circuits. But this IC is cheap and widely available, so if your application doesn't demand super high-grade performance, the 741 will usually do a fine job.

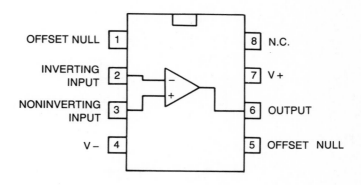

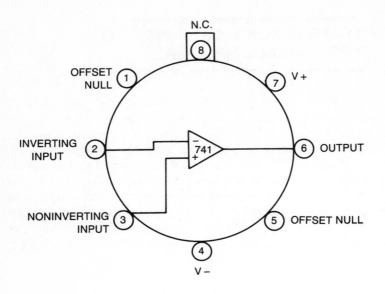

N.C.

OFFSET NULL ① 1

INVERTING INPUT ② 2 741

NONINVERTING INPUT ③ 3

④ 4

V −

⑦ 7 V +

⑥ 6 OUTPUT

⑤ 5 OFFSET NULL

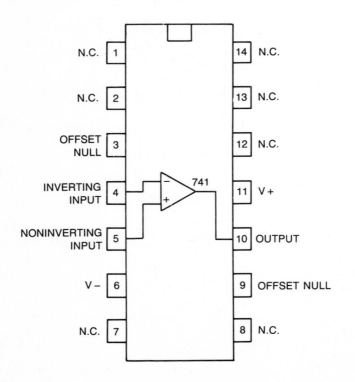

N.C. 1

N.C. 2

OFFSET NULL 3

INVERTING INPUT 4

NONINVERTING INPUT 5

V − 6

N.C. 7

14 N.C.

13 N.C.

12 N.C.

741 11 V +

10 OUTPUT

9 OFFSET NULL

8 N.C.

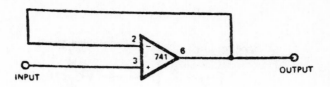

Unity-gain voltage follower

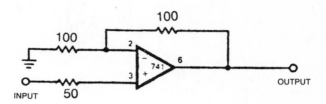

Non-inverting amplifier

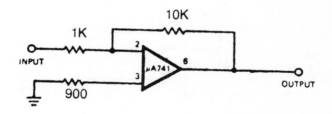

Inverting amplifier

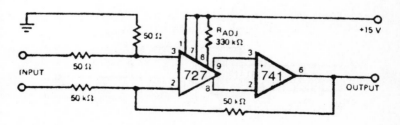

Low drift low noise amplifier

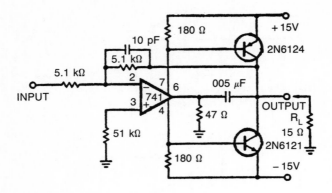

High slew rate power amplifier

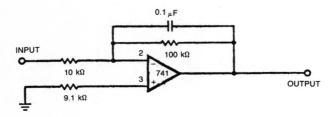

Simple integrator

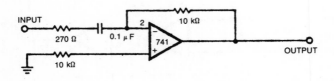

Simple differentiator

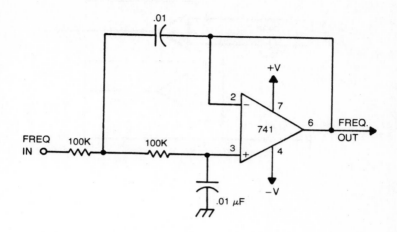

Low-pass active filter

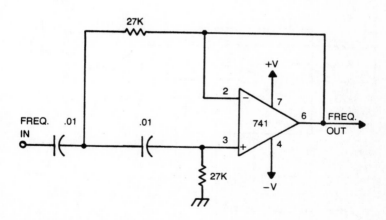

High-pass active filter

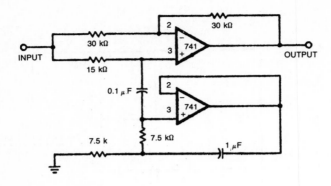

Notch filter

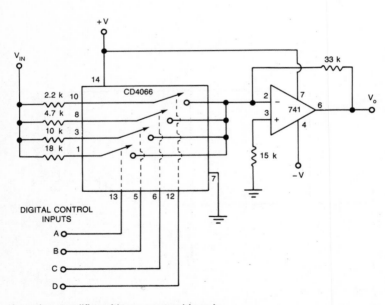

Inverting amplifier with programmable gain

98

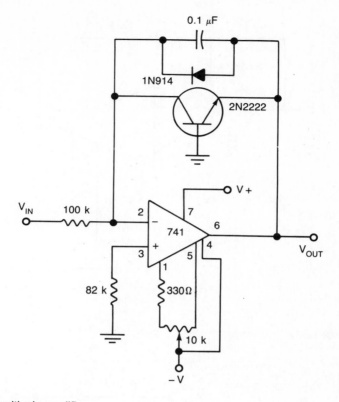

Logarithmic amplifier

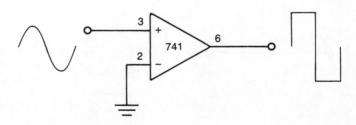

Sine wave to square-wave converter

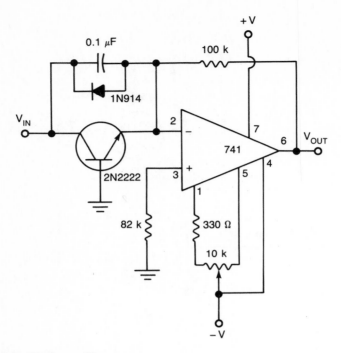

Antilogarithmic amplifier

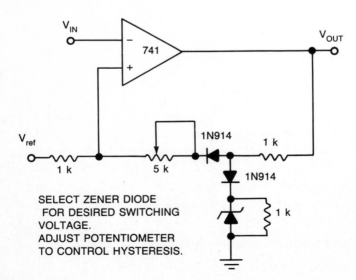

SELECT ZENER DIODE
FOR DESIRED SWITCHING
VOLTAGE.
ADJUST POTENTIOMETER
TO CONTROL HYSTERESIS.

Schmitt trigger

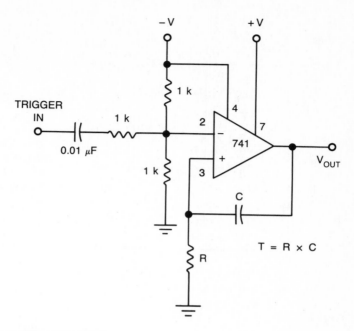

Monostable multivibrator

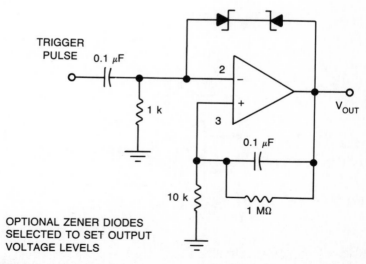

OPTIONAL ZENER DIODES
SELECTED TO SET OUTPUT
VOLTAGE LEVELS

Bistable multivibrator

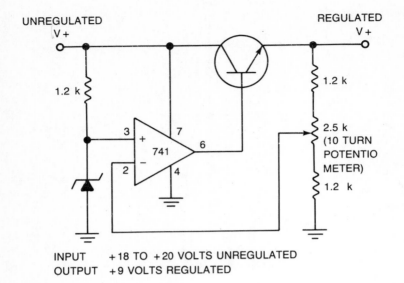

INPUT +18 TO +20 VOLTS UNREGULATED
OUTPUT +9 VOLTS REGULATED

Voltage regulator

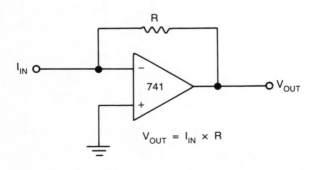

$V_{OUT} = I_{IN} \times R$

Current to voltage converter

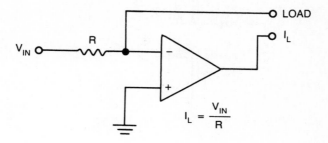

Voltage to current converter

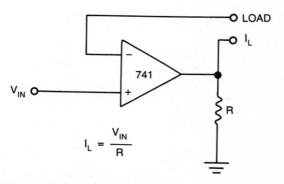

Noninverting voltage to current converter

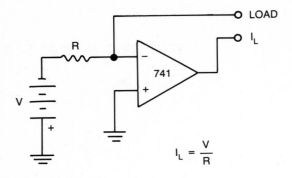

$$I_L = \frac{V}{R}$$

Constant current source

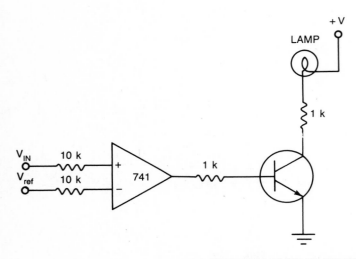

TRANSISTOR SHOULD BE SELECTED
TO SUPPLY SUFFICIENT CURRENT
TO LOAD LAMP

Comparator lamp

104

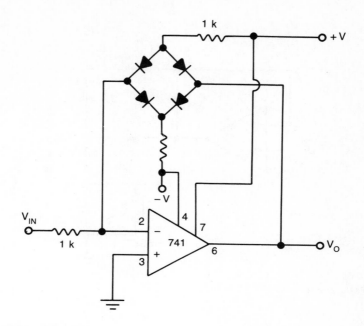

Threshold detector

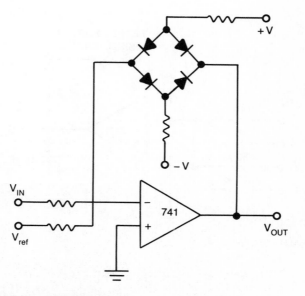

Nonsymmetrical threshold detector

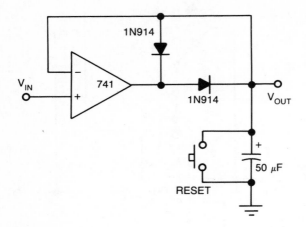

Peak detector

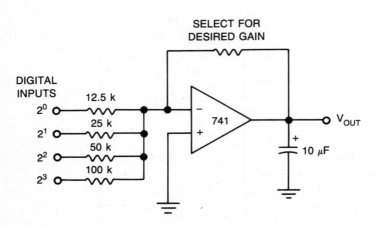

Simple 4-bit D/A converter

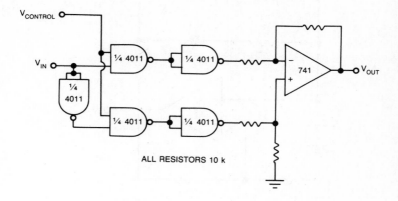

OP amp logic level shifter

$$F = \frac{5}{RC}$$

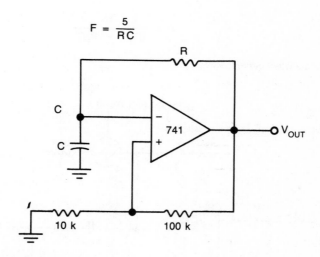

Square-wave generator

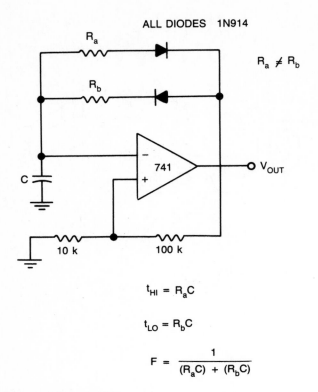

ALL DIODES 1N914

$R_a \neq R_b$

$$t_{HI} = R_a C$$

$$t_{LO} = R_b C$$

$$F = \frac{1}{(R_a C) + (R_b C)}$$

Rectangle-wave generator

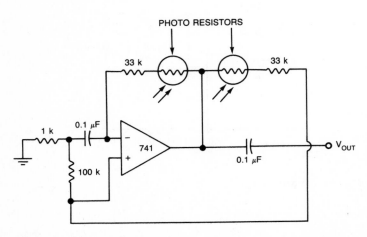

PHOTO RESISTORS

Light sensitive tone generator

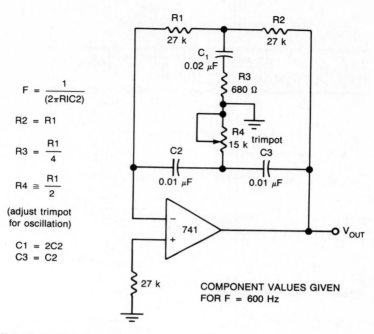

$$F = \frac{1}{(2\pi R1C2)}$$

$$R2 = R1$$

$$R3 = \frac{R1}{4}$$

$$R4 \cong \frac{R1}{2}$$

(adjust trimpot for oscillation)

$$C1 = 2C2$$
$$C3 = C2$$

COMPONENT VALUES GIVEN FOR F = 600 Hz

Twin-T sine wave oscillator

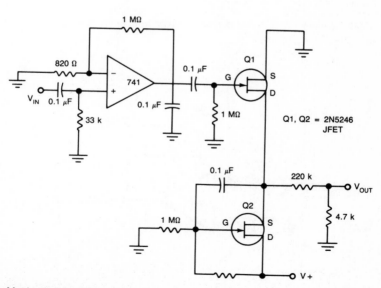

Musical fuzz effect

109

747—DUAL FREQUENCY COMPENSATED OPERATIONAL AMPLIFIER

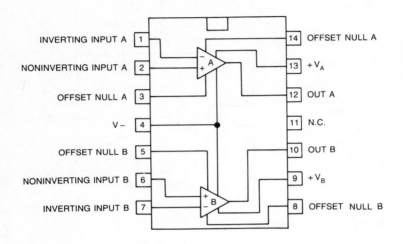

Contains two independent 741 type op amps in a single package

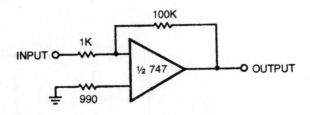

Inverting amplifier

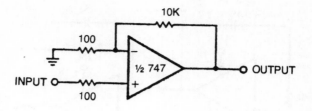

Noninverting amplifier

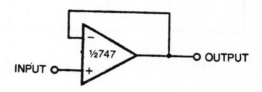

Unity-gain voltage follower

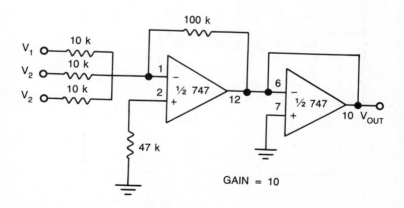

GAIN = 10

Noninverting summing amplifier

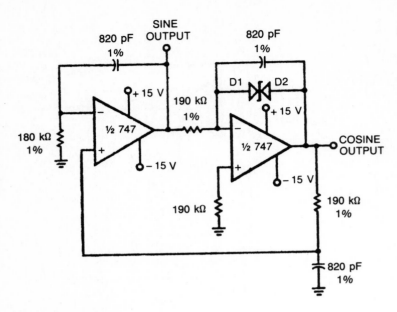

Quadrature oscillator

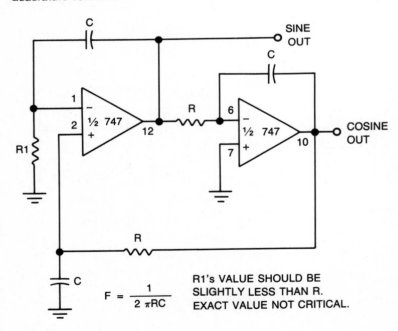

$$F = \frac{1}{2\pi RC}$$

R1's VALUE SHOULD BE SLIGHTLY LESS THAN R. EXACT VALUE NOT CRITICAL.

Alternate quadrature oscillator

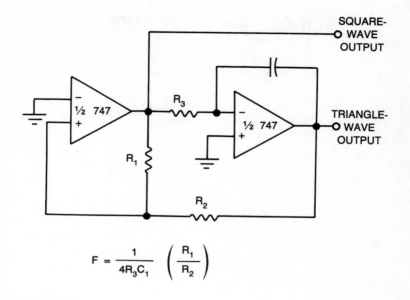

$$F = \frac{1}{4R_3C_1} \left(\frac{R_1}{R_2} \right)$$

Function generator

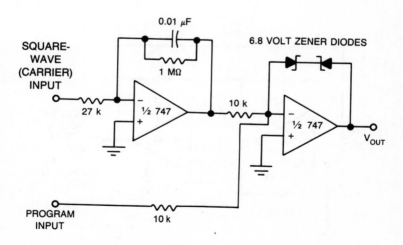

113

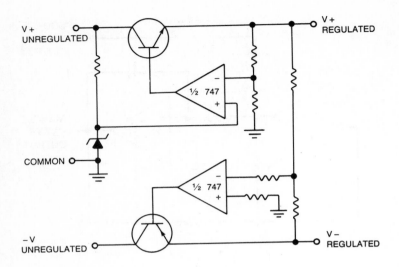

Dual polarity power supply

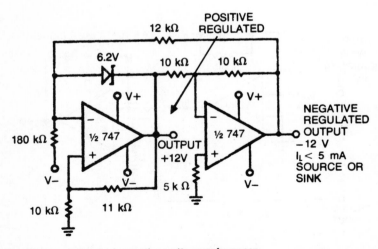

Tracking positive and negative voltage references

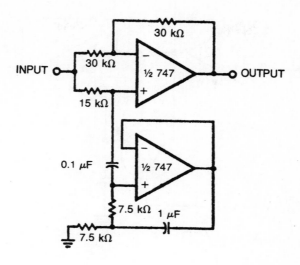

Notch filter

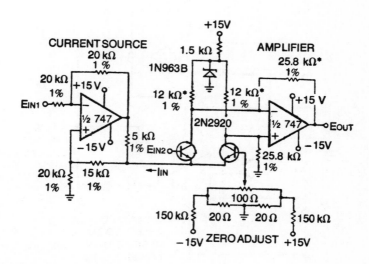

Analog multiplier

115

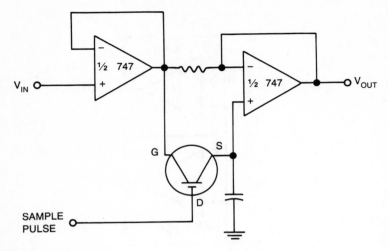

Sample and hold

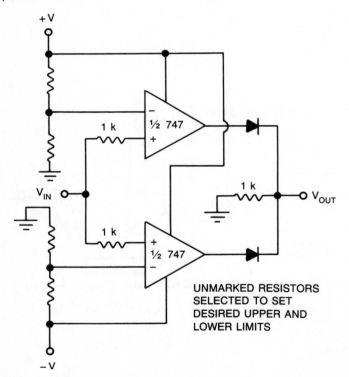

UNMARKED RESISTORS
SELECTED TO SET
DESIRED UPPER AND
LOWER LIMITS

Window comparator

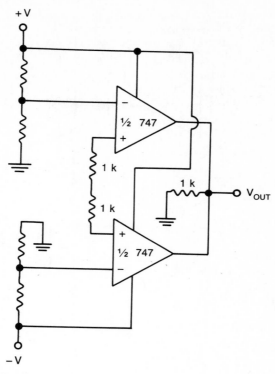

Magnitude detector

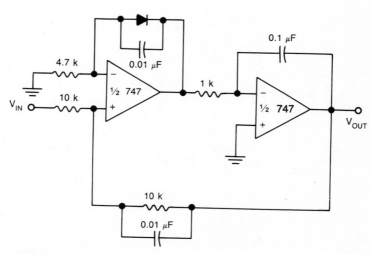

Inverting peak detector

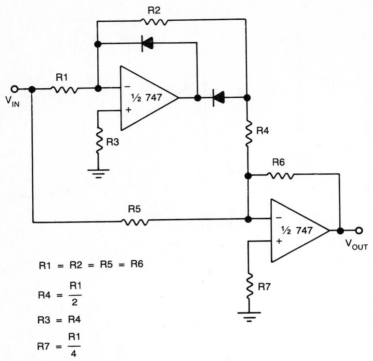

$R1 = R2 = R5 = R6$

$R4 = \dfrac{R1}{2}$

$R3 = R4$

$R7 = \dfrac{R1}{4}$

Absolute value extractor

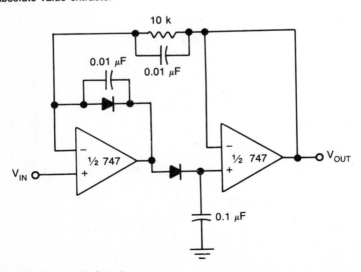

Noninverting peak detector

118

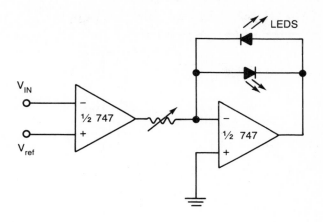

Null indicator

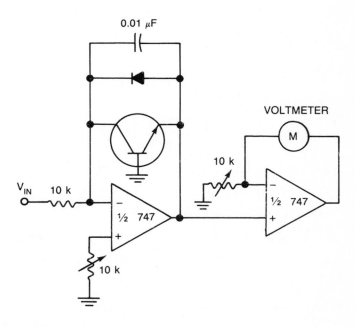

Decibel meter

748—HIGH PERFORMANCE OPERATIONAL AMPLIFIER

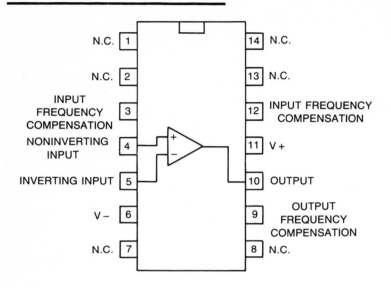

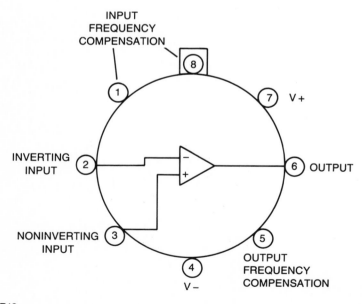

748

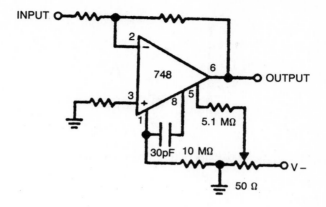

Inverting amplifier with balancing circuit

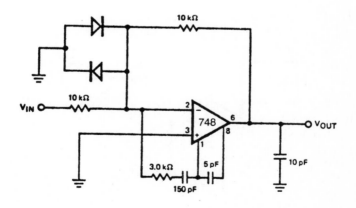

Feed forward compensation

121

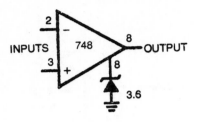

Voltage comparator for driving DTL or TTL integrated circuits

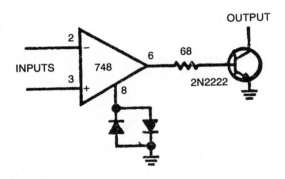

Voltage comparator for driving RTL logic or high current driver

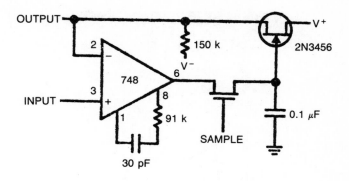

Low drift sample and hold

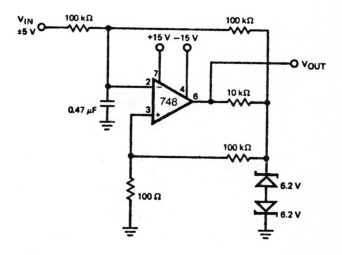

Pulse width modulator

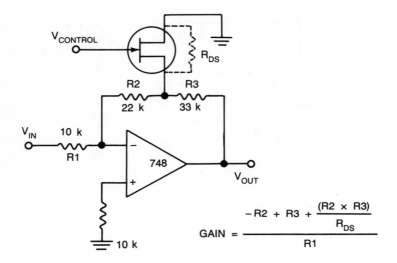

$$\text{GAIN} = \frac{-R2 + R3 + \dfrac{(R2 \times R3)}{R_{DS}}}{R1}$$

AGC amplifier

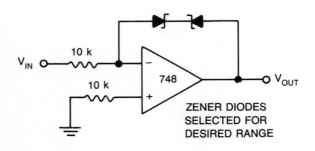

ZENER DIODES
SELECTED FOR
DESIRED RANGE

Zero crossing detector

124

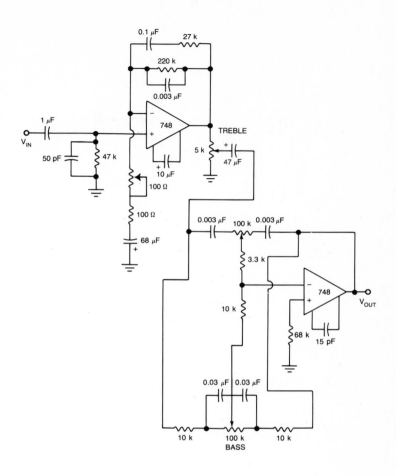

Audio preamp with tone controls

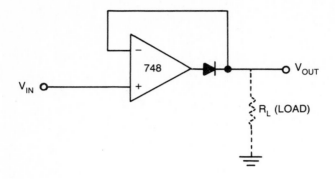

Precision diode

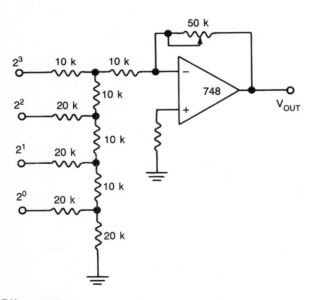

R-2R D/A converter

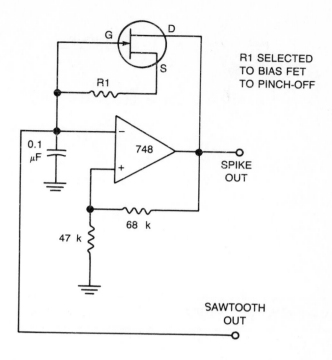

Sawtooth-wave generator

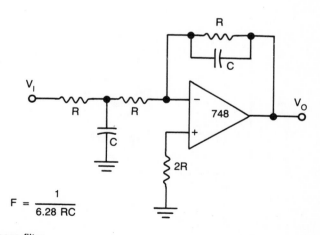

$$F = \frac{1}{6.28 \; RC}$$

Low-pass filter

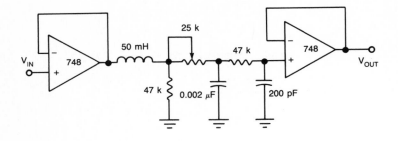

Record scratch filter

791—POWER OPERATIONAL AMPLIFIER

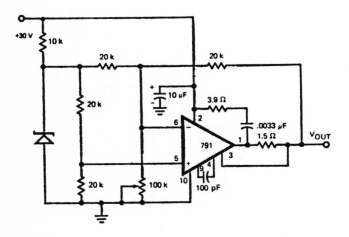

Positive voltage regulator

128

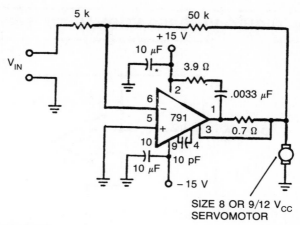

DC servo amplifier

798—DUAL OPERATIONAL AMPLIFIER

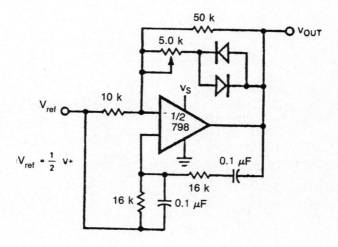

Wien bridge oscillator

129

799—HIGH GAIN OPERATIONAL AMPLIFIER

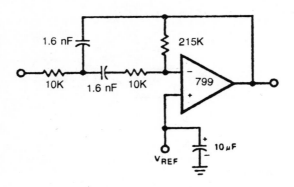

Multiple feedback bandpass filter

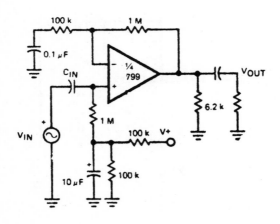

AC couple noninverting amplifier

130

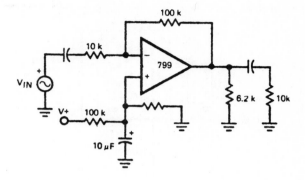

AC coupled inverting amplifier

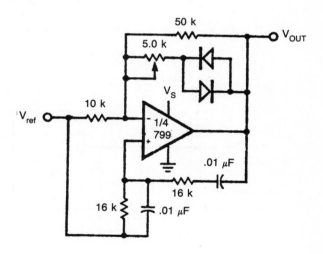

Wien bridge oscillator

131

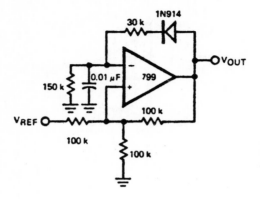

Pulse generator

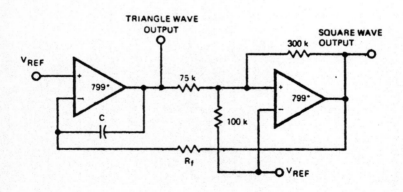

Function generator

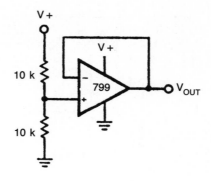

Voltage reference

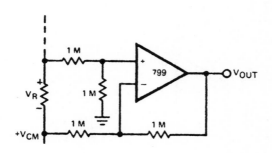

Ground referencing a differential input signal

0001—LOW POWER OPERATIONAL AMPLIFIER

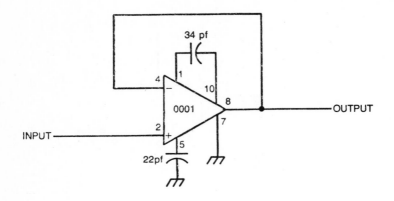

Voltage follower

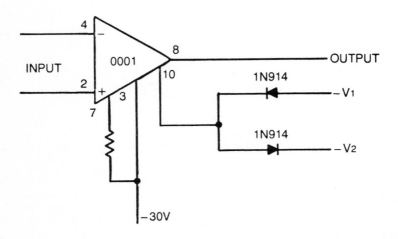

Voltage comparator and MOS driver

0003—WIDE BANDWIDTH OPERATIONAL AMPLIFIER

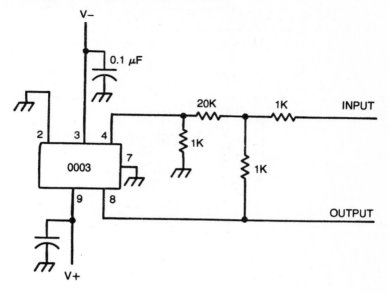

Unity gain inverting amplifier

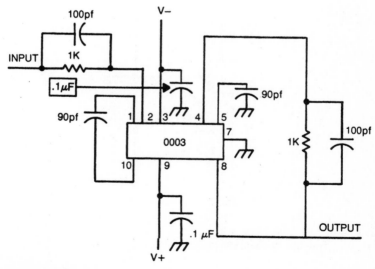

Unity gain follower

0004—HIGH VOLTAGE OPERATIONAL AMPLIFIER

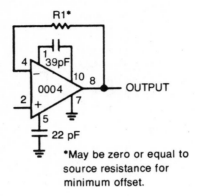

*May be zero or equal to source resistance for minimum offset.

Voltage follower

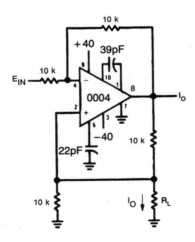

High compliance current source

0005—HIGH INPUT RESISTANCE
OPERATIONAL AMPLIFIER

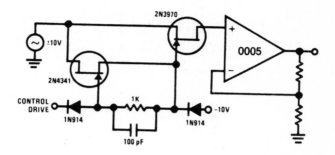

High toggle rate high frequency analog switch

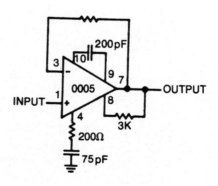

Voltage follower

0021—1.0 AMP OPERATIONAL AMPLIFIER

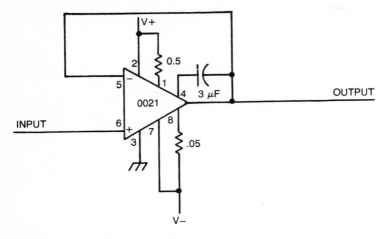

Unity gain amplifier with short circuit limiting

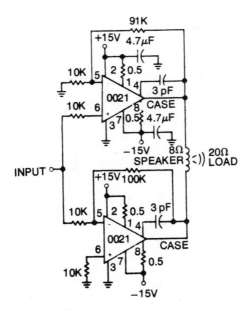

10 watt (rms) audio amplifier

0022–FET INPUT OPERATIONAL AMPLIFIER

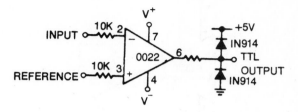

Precision voltage comparator

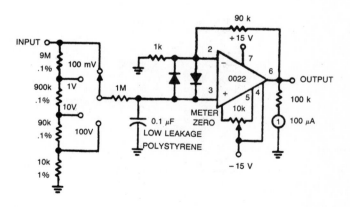

Sensitive low cost "VTVM"

0024—HIGH SLEW RATE OPERATIONAL AMPLIFIER

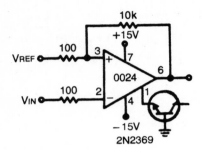

TTL compatible comparator

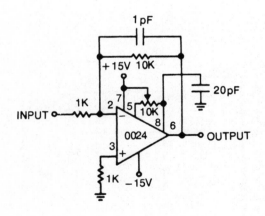

Offset null

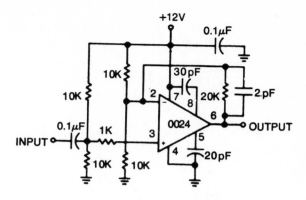

Video amplifier

0032—ULTRA FAST FET OPERATIONAL AMPLIFIER

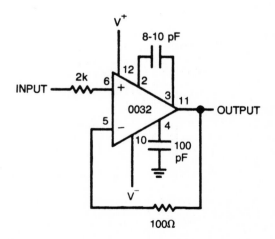

Unity gain amplifier

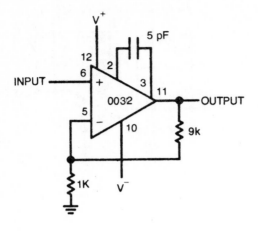

10X buffer amplifier

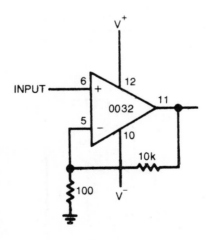

100X buffer amplifier

142

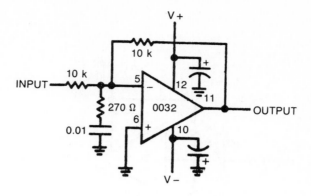

Noncompensated unity gain inverter

0033—FAST BUFFER AMPLIFIER

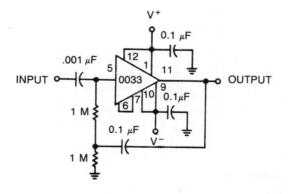

High input impedance ac coupled amplifier

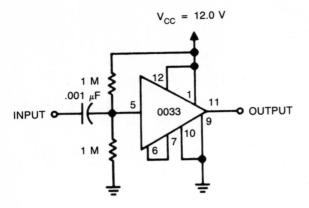

Single supply ac amplifier

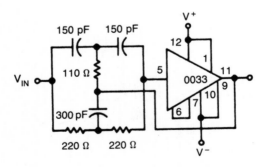

4.5 MHz notch filter

0042—FET INPUT OPERATIONAL AMPLIFIER

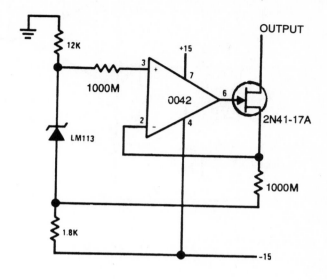

Precision current sink

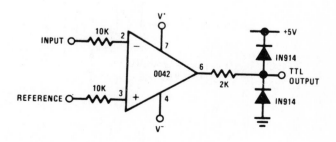

Precision voltage comparator

145

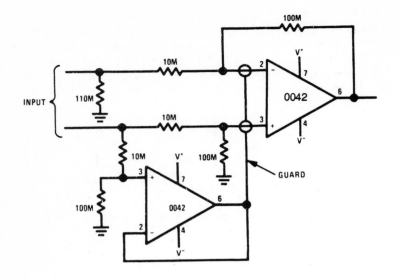

Guarded full differential amplifier

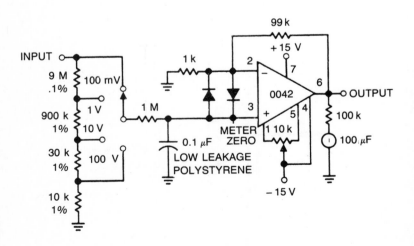

Sensitive low cost "VTVM"

146

0044—LOW NOISE OPERATIONAL AMPLIFIER

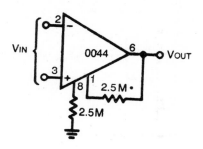

X1000 Instrumentation amp

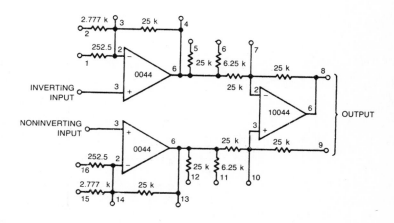

Precision instrumentation amplifier

147

0062—HIGH SPEED FET OPERATIONAL AMPLIFIER

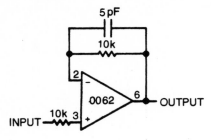

Fast voltage follower

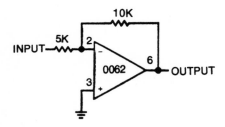

Fast summing amplifier

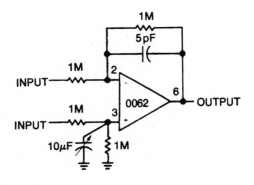

Differential amplifier

148

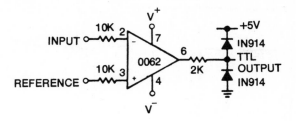

Fast precision voltage comparator

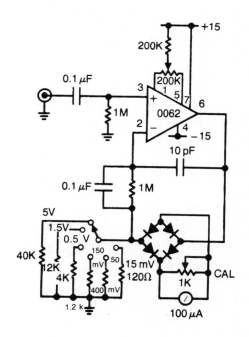

Wide range ac voltmeter

1458—DUAL COMPENSATED OPERATIONAL AMPLIFIER

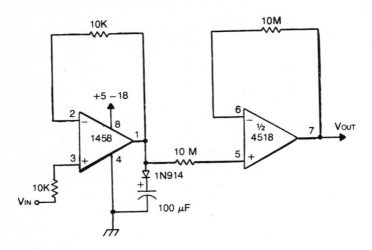

Peak detector

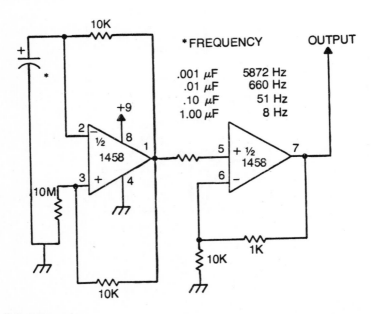

Pulse generator

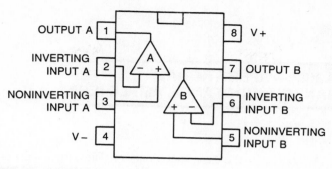

OUTPUT A
INVERTING INPUT A
NONINVERTING INPUT A
V −

V +
OUTPUT B
INVERTING INPUT B
NONINVERTING INPUT B

COMPATIBLE WITH ANY 741
APPLICATION THAT DOES NOT
REQUIRE THE OFFSET NULL PINS

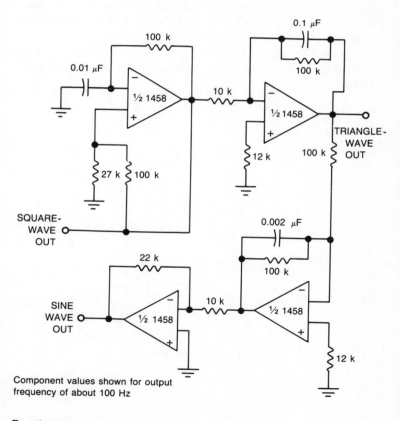

Component values shown for output
frequency of about 100 Hz

Function generator

151

CA3080—OPERATIONAL TRANSCONDUCTANCE AMPLIFIER

A variation on the basic operational amplifier, or op amp, is the Operational Transconductance Amplifier, or OTA. The CA3080 is a typical IC of this type.

Generally, op amps are intended to be used as voltage amplifiers. An OTA, on the other hand, is a voltage-to-current amplifier. This difference is indicated in schematic diagrams by adding a constant current symbol (two interlocked circles) to the output of an op amp symbol.

The OTA can be used in many of the ordinary op amp applications. It can also be employed in some specialized applications of its own.

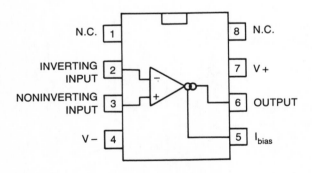

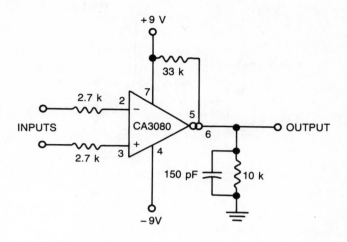

Differential amplifier

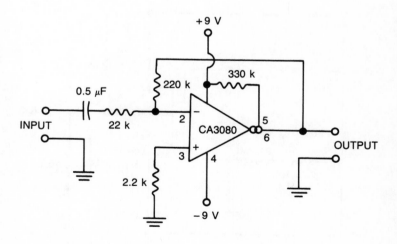

Inverting amplifier with low power consumption

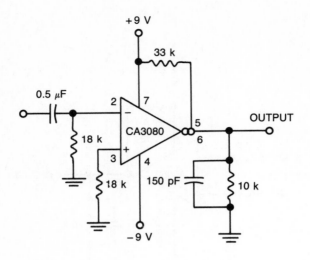

AC coupled inverting amplifier

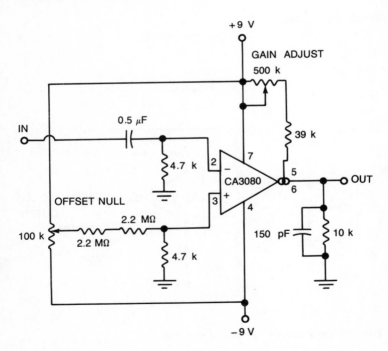

AC amplifier with variable gain

154

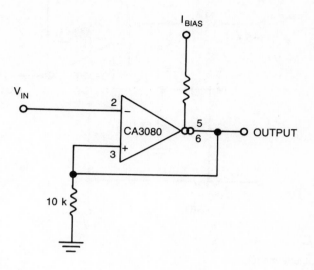

Schmitt trigger

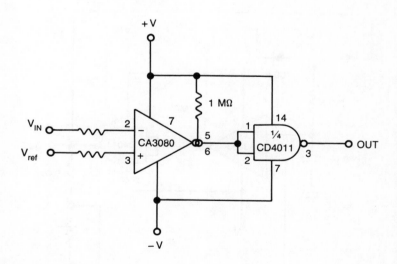

Low power comparator

155

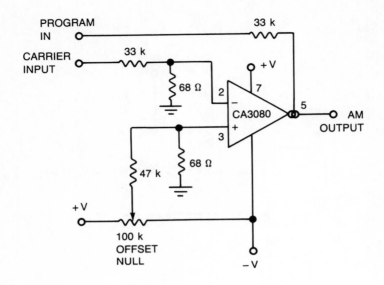

Amplitude modulator

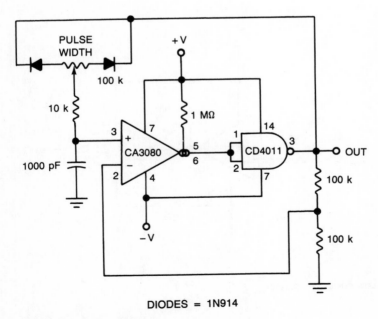

DIODES = 1N914

Variable width pulse generator

156

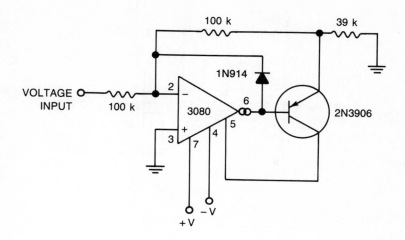

Precision current source

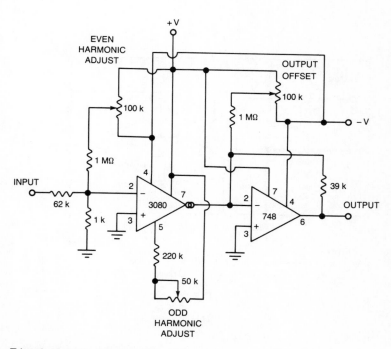

Triangle wave to sine-wave converter

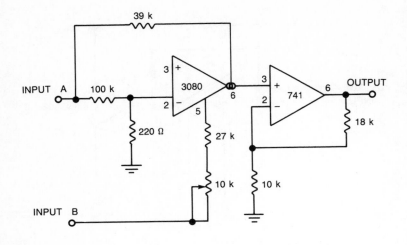

Four quadrant multiplier

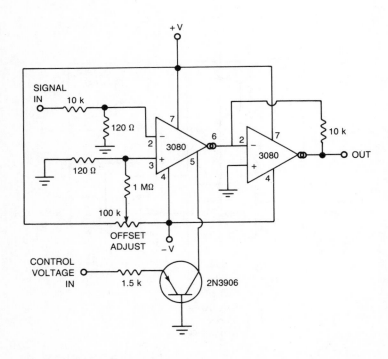

VCA (voltage-controlled amplifier)

158

3301—QUAD SINGLE SUPPLY OPERATIONAL AMPLIFIER

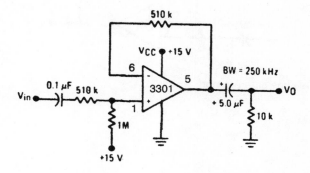

Noninverting amplifier

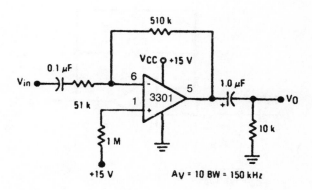

Inverting amplifier

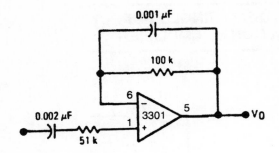

Positive-edge differentiator

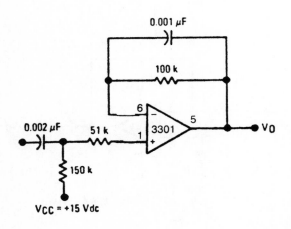

Negative-edge differentiator

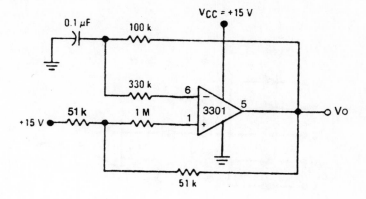

Astable multivibrator

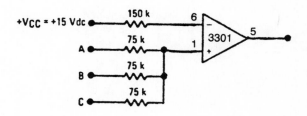

Logic OR gate

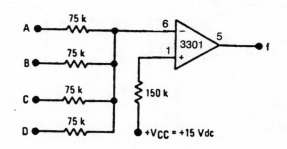

Logic NOR gate

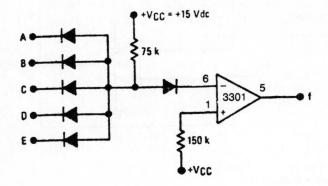

Logic NAND gate (Large Fan-In)

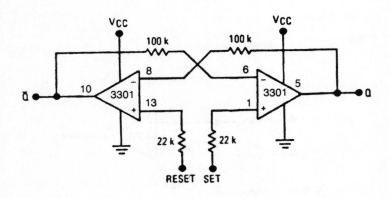

R-S flip-flop

162

LM 3900—QUAD NORTON AMPLIFIER

Applications for Norton amplifiers are similar to those for ordinary op amps, except the output voltage responds to input currents rather than input voltages.

The LM3900 contains four independent Norton amplifier stages in a single 14 pin package. The amplifiers are internally frequency compensated. External frequency compensation components are not needed.

The LM3900 can be powered from a dual polarity supply from ±2 volts to ±18 volts, or it can be driven by a single polarity supply voltage in the +4 to +36 volt range.

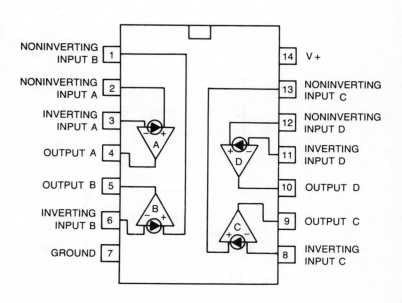

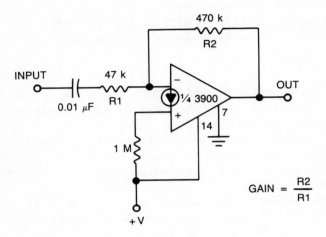

GAIN = $\dfrac{R2}{R1}$

Inverting amplifier

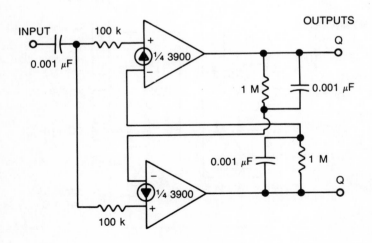

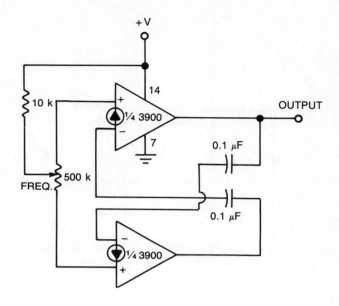

Square-wave generator

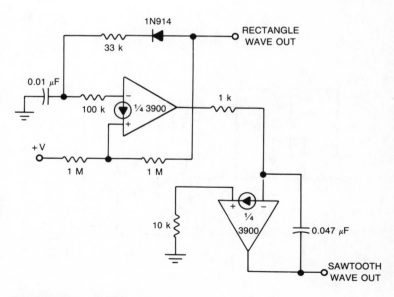

Function generator

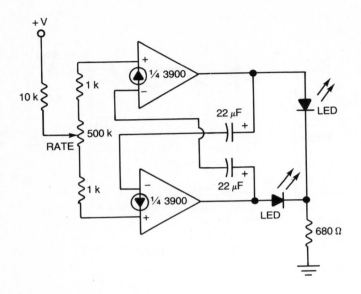

Dual LED flasher

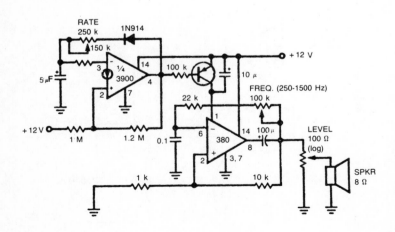

Siren with programmable frequency and rate adjustment

4131—HIGH GAIN OPERATIONAL AMPLIFIER

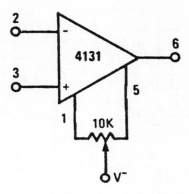

Voltage offset null circuit

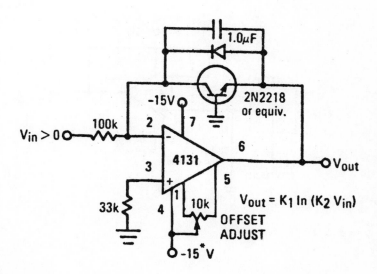

$$V_{out} = K_1 \ln (K_2 V_{in})$$

Logarithmic amplifier

167

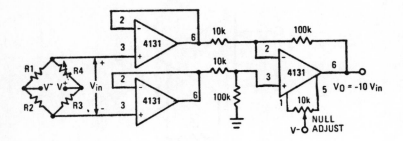

High impedance bridge amplifier

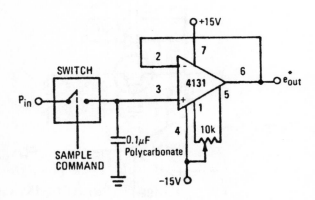

Low drift sample and hold

4132—MICROPOWER OPERATIONAL AMPLIFIER

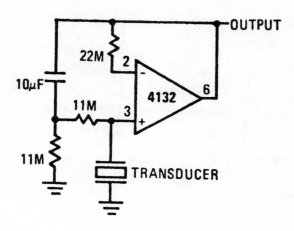

Amplifier for Piezoelectric transducers

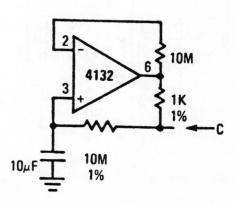

Capacitance multiplier

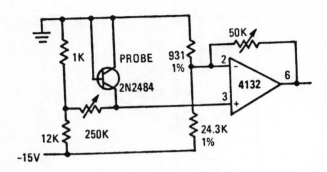

Temperature probe

4136—QUAD 741 OPERATIONAL AMPLIFIER

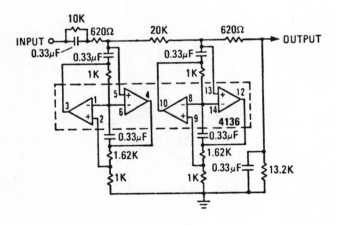

400 Hz low-pass Butterworth active filter

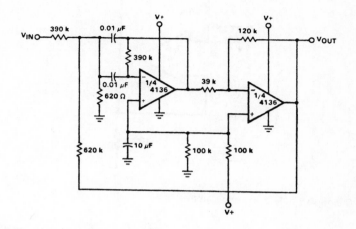

1 kHz bandpass active filter

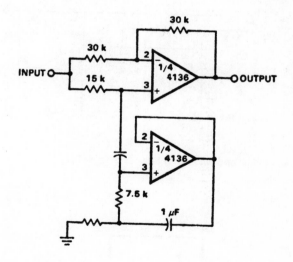

Notch filter

171

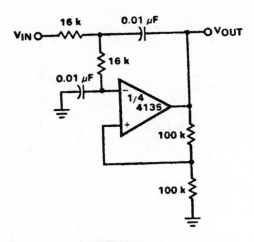

DC coupled 1 kHz low-pass active filter

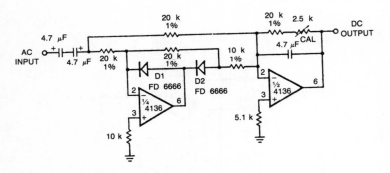

Full-wave rectifier and averaging filter

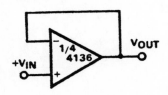

Voltage follower

172

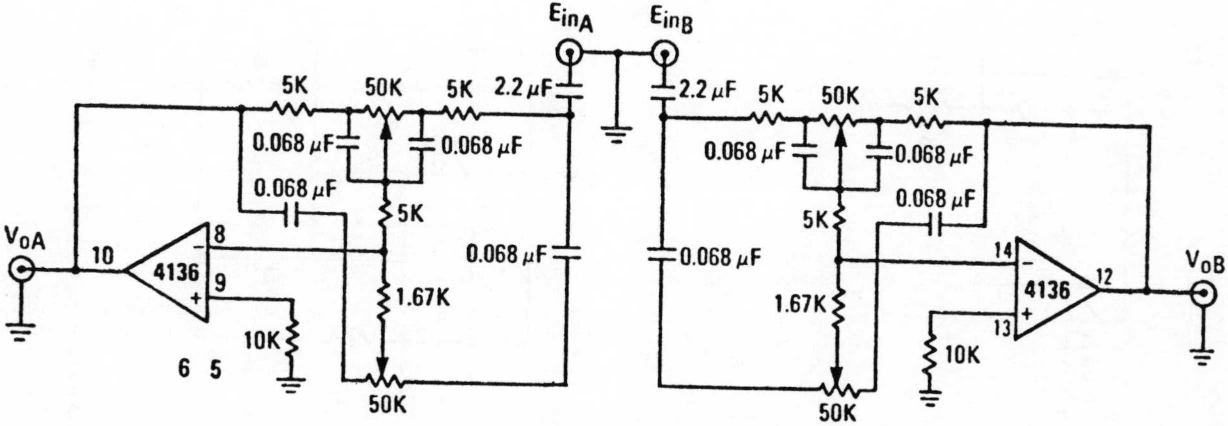

Stereo tone control

173

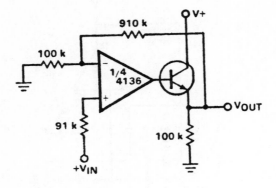

Power amplifier

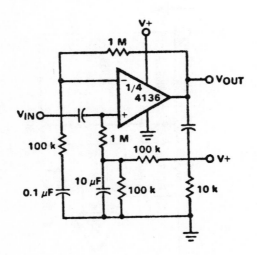

AC coupled noninverting amplifier

174

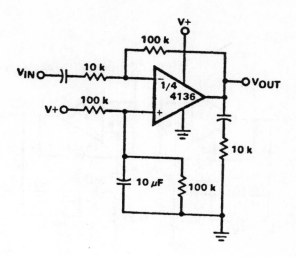

AC coupled inverting amplifier

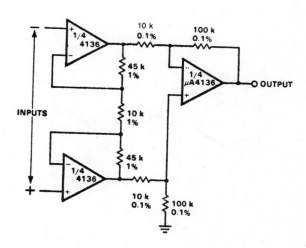

Differential input instrumentation amplifier with high common mode rejection

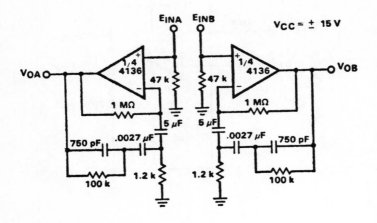

RIAA preamplifier

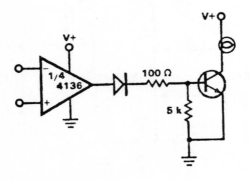

Lamp driver

176

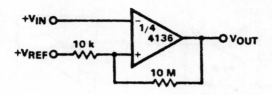

Comparator with hysteresis

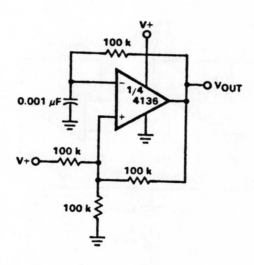

Square-wave oscillator

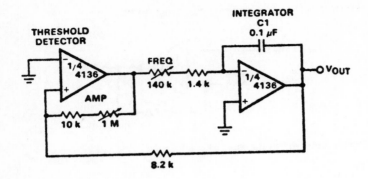

Triangular-wave generator

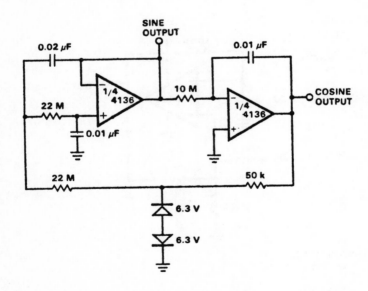

Low frequency sine-wave generator with quadrature output

4250—PROGRAMMABLE OPERATIONAL AMPLIFIER

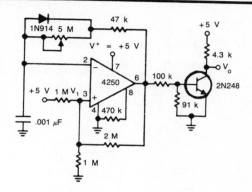

Pulse generator

TL084C QUAD OPERATIONAL AMPLIFIER WITH JFET INPUTS

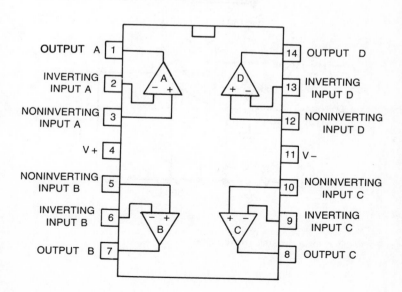

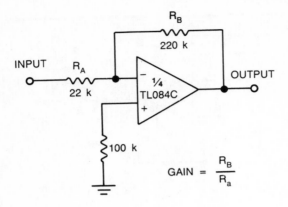

Inverting dc amplifier

$$GAIN = \frac{R_B}{R_a}$$

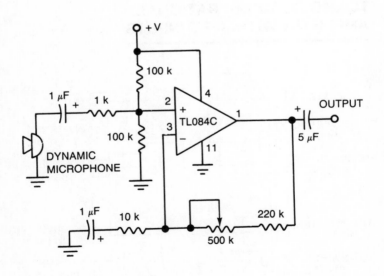

Preamplifier for a dynamic microphone

180

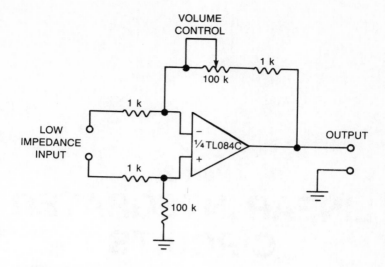

Low input impedance preamplifier

Part 2
LINEAR INTEGRATED CIRCUITS

While op amps are probably the most popular and most versatile class of linear integrated circuits, there are many other types. Some are dedicated devices for applications similar to those using op amps. This category would include comparators and amplifiers. Other linear ICs, such as timers, are intended for entirely different applications.

311—VOLTAGE COMPARATOR

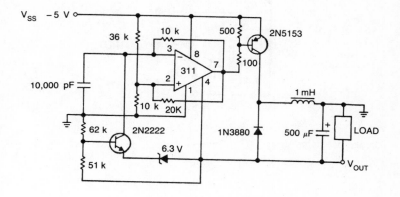

Switching regulator for voltage conversion

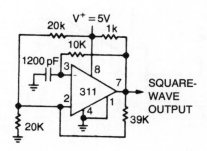

100 kHz free-running multivibrator

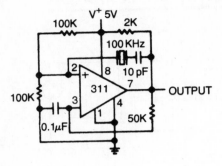

Crystal oscillator

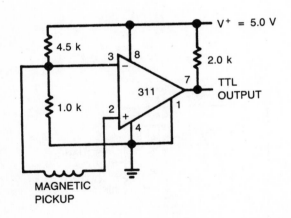

Detector for magnetic transducer

322—PRECISION TIMER

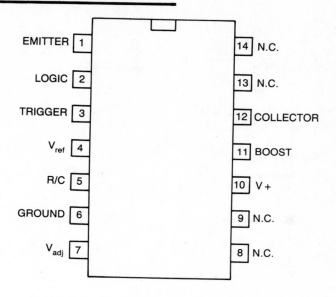

EMITTER	1		14	N.C.
LOGIC	2		13	N.C.
TRIGGER	3		12	COLLECTOR
V_{ref}	4		11	BOOST
R/C	5		10	V +
GROUND	6		9	N.C.
V_{adj}	7		8	N.C.

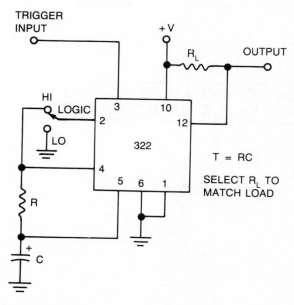

$T = RC$

SELECT R_L TO
MATCH LOAD

Monostable multivibrator

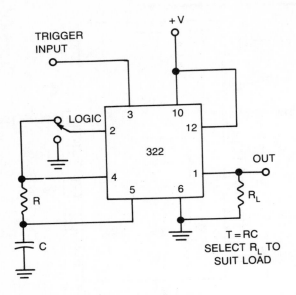

Monostable multivibrator with common-emitter output

LM334—TEMPERATURE SENSOR/ ADJUSTABLE CURRENT SOURCE

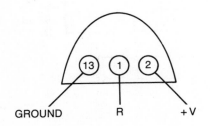

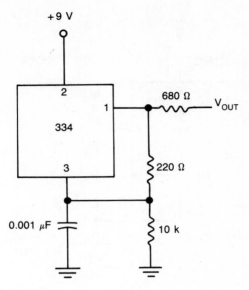

Simple thermometer

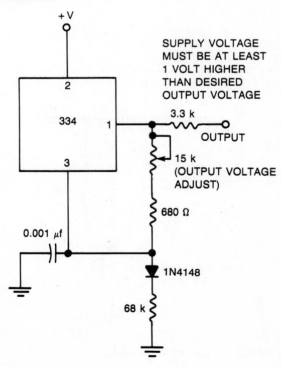

SUPPLY VOLTAGE MUST BE AT LEAST 1 VOLT HIGHER THAN DESIRED OUTPUT VOLTAGE

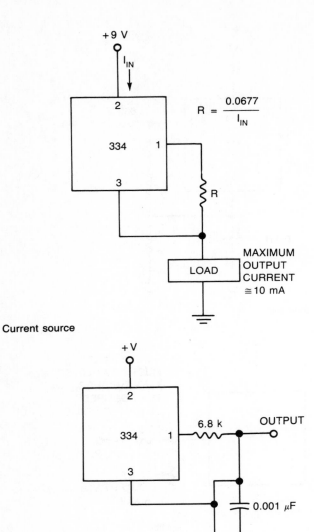

$$R = \frac{0.0677}{I_{IN}}$$

Current source

Rectangle wave to sawtooth wave converter

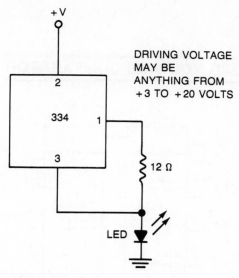

DRIVING VOLTAGE
MAY BE
ANYTHING FROM
+3 TO +20 VOLTS

Constant-output LED driver

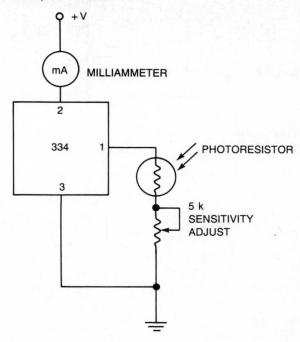

Simple light meter

339—QUAD VOLTAGE COMPARATOR

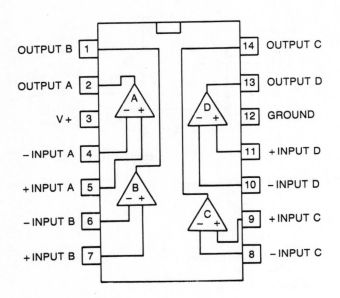

One-shot multivibrator

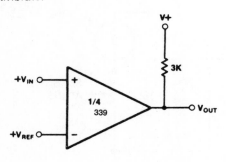

Basic comparator

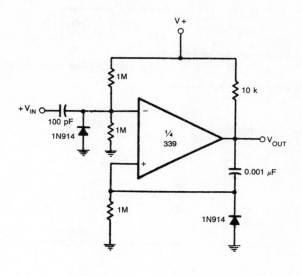

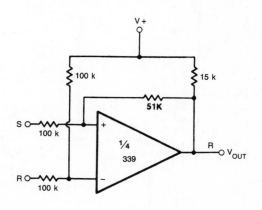

Bi-stable multivibrator

191

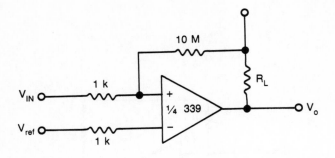

Comparator with hysteresis

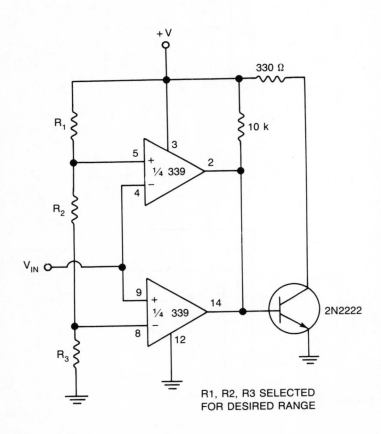

R1, R2, R3 SELECTED
FOR DESIRED RANGE

Limit comparator

192

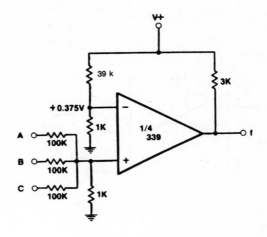

AND gate

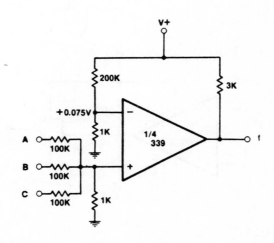

OR gate

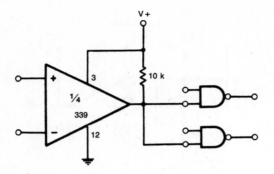

Driving TTL

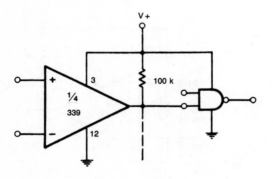

Driving CMOS

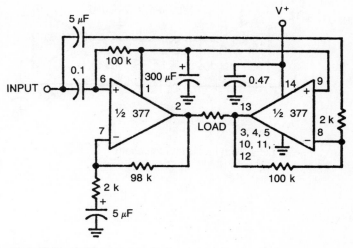

4-watt bridge amplifier

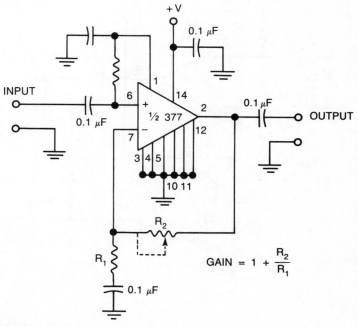

$$GAIN = 1 + \frac{R_2}{R_1}$$

Noninverting amplifier

195

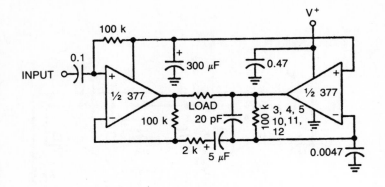

4-watt bridge amplifier with high input impedance

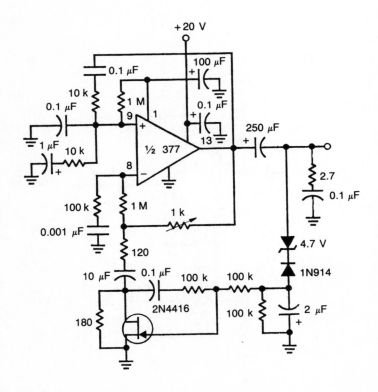

Wien bridge power oscillator

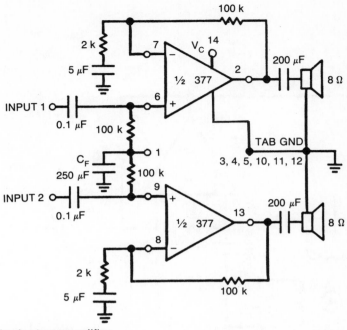

Simple stereo amplifier

377—DUAL TWO WATT AUDIO AMPLIFIER

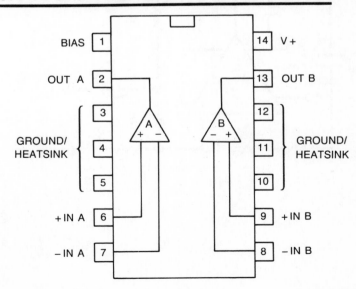

Two-phase motor drive

378—DUAL FOUR-WATT AUDIO AMPLIFIER

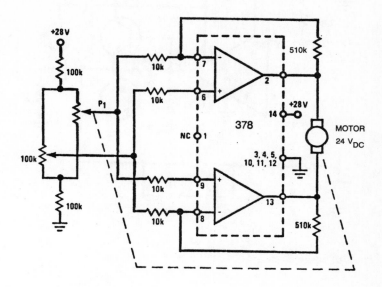

Proportional speed controller

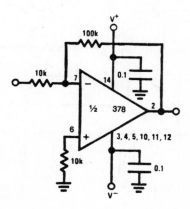

Power op amp (using split supplies)

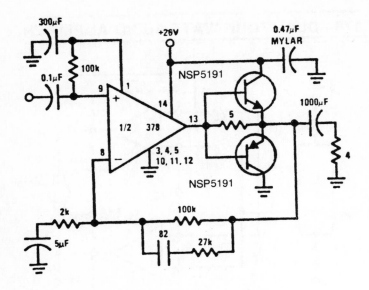

10-Watt power amplifier

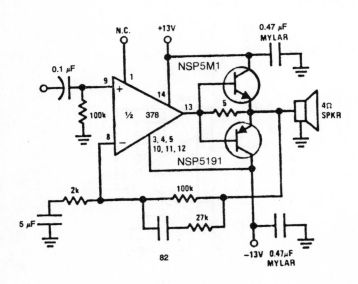

12-Watt low-distortion power amplifier

379—DUAL SIX-WATT AUDIO AMPLIFIER

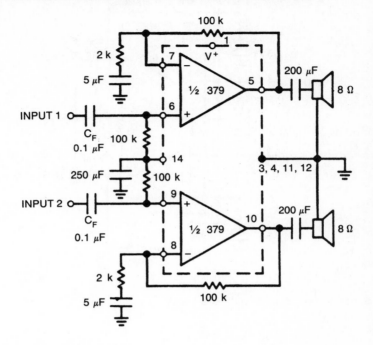

Simple stereo amplifier

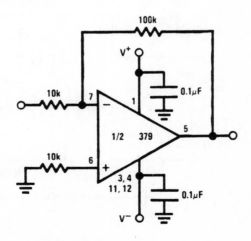

Power op amp (using split supplies)

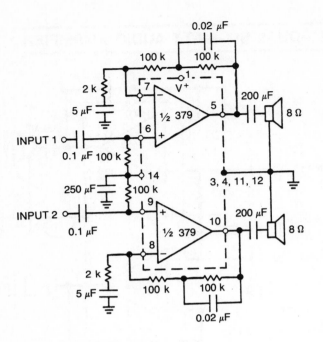

Simple stereo amplifier with bass boost

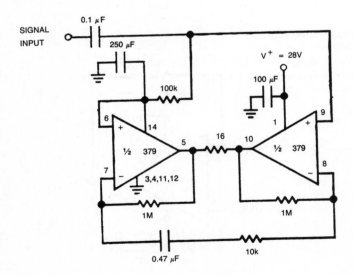

12W bridge amplifier

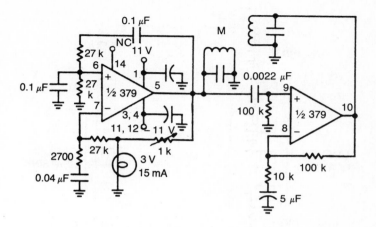

Two-phase motor drive

380—AUDIO POWER AMPLIFIER

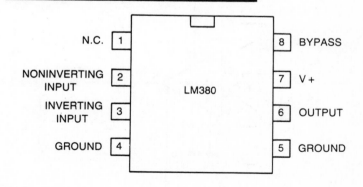

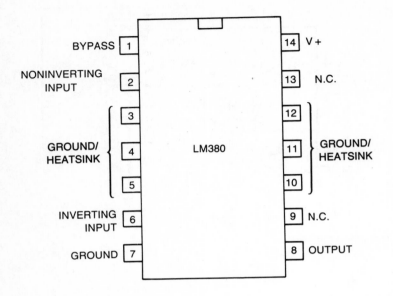

LM380—Audio Amplifier

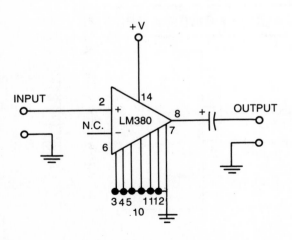

BASic audio amplifier

204

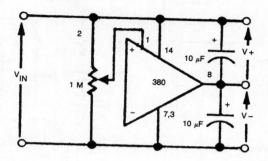

Dual supply

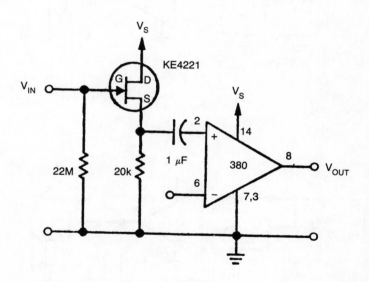

High input impedance

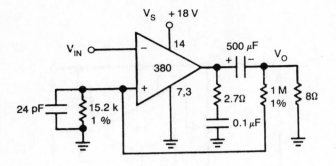

Boosted gain of 200 using positive feedback

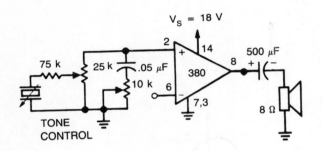

Phono amp

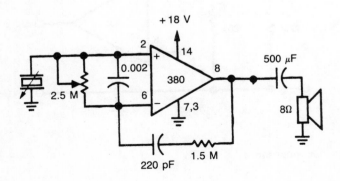

RIAA phono amplifier

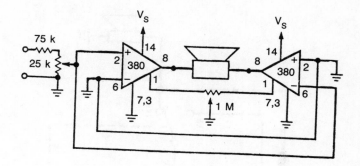

Voltage divider input

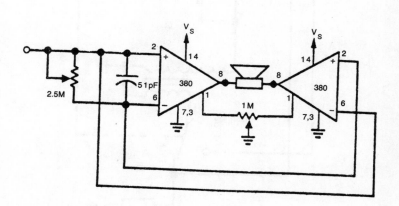

Quiescent balance control

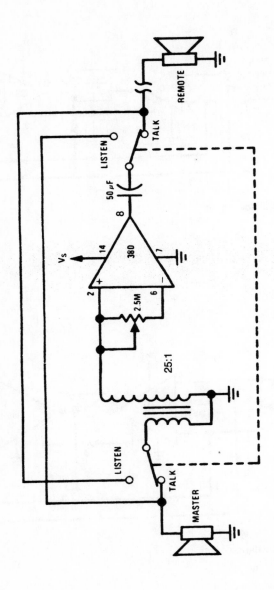

Intercom

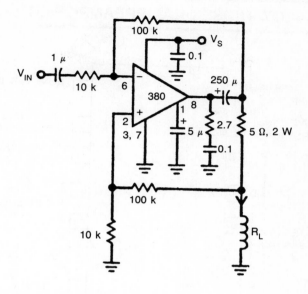

Power voltage-to-current converter

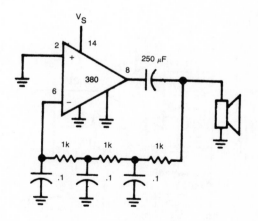

Phase shift oscillator

381—LOW-NOISE DUAL PREAMPLIFIER

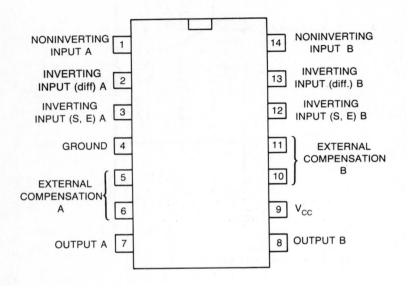

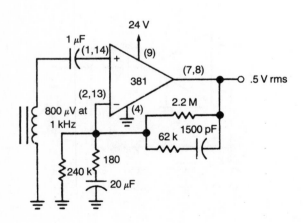

Tape playback amplifier

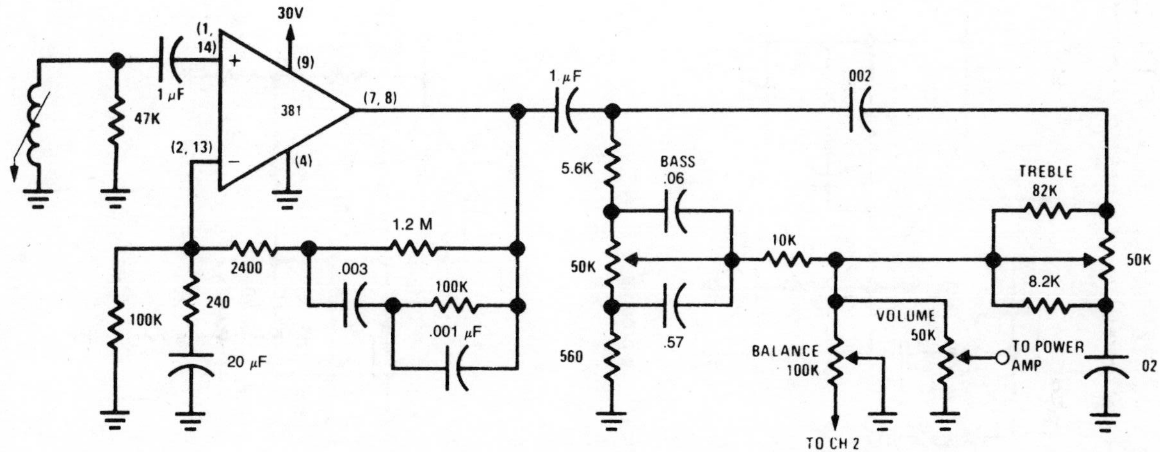

Single channel of complete phono preamp

211

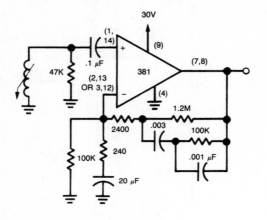

Typical magnetic phono preamp

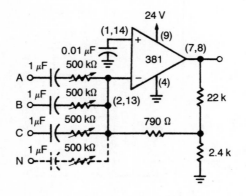

Audio mixer

212

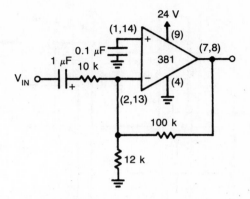

Ultra-low distortion amplifier

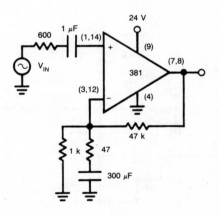

Flat gain circuit

381A—LOW-NOISE DUAL PREAMPLIFIER

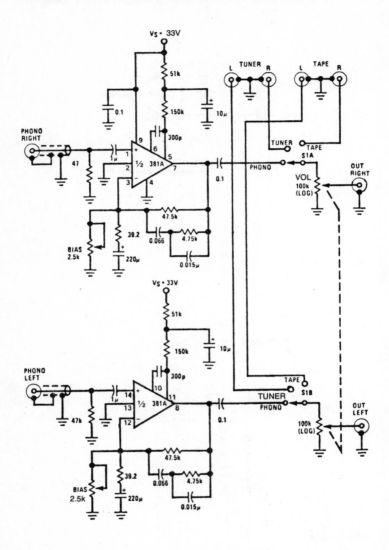

Ultra-low noise mini preamp (RIAA)

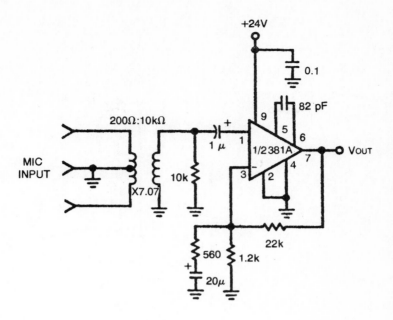

Balanced input mic preamp

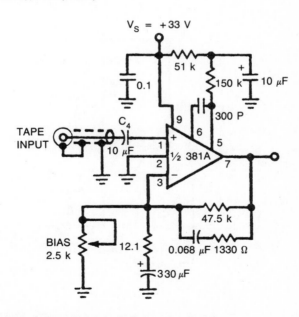

Ultra-low-noise tape preamp (NAB, 1-7/8 & 3-3/4 IPS)

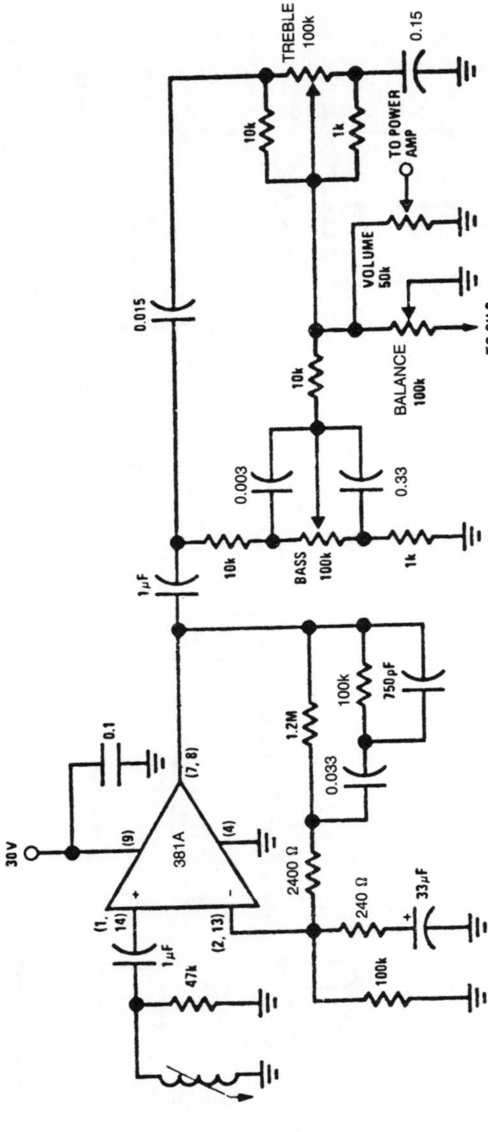

Single channel of complete phono preamp

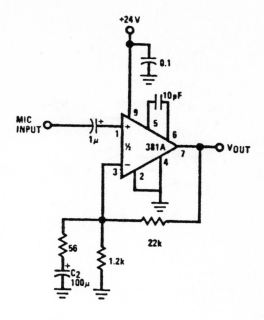

Unbalanced input mic preamp

382—LOW-NOISE DUAL PREAMPLIFIER

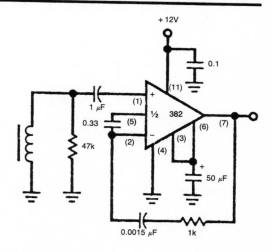

Phono preamp (RIAA)

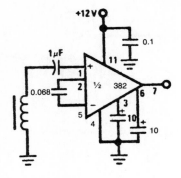

Tape preamp (NAB, 1-7/8 & 3-3/4 IPS)

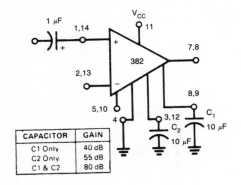

CAPACITOR	GAIN
C1 Only	40 dB
C2 Only	55 dB
C1 & C2	80 dB

Flat response — fixed gain configuration

384—FIVE-WATT AUDIO POWER AMPLIFIER

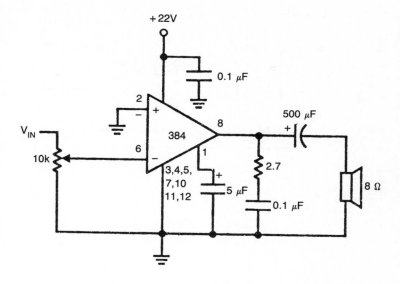

Typical 5W amplifier

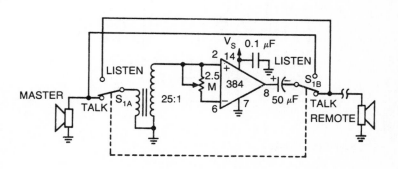

Intercom

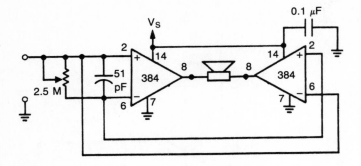

Bridge amplifier

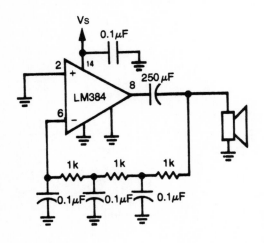

Phase shift oscillator

386—LOW VOLTAGE
AUDIO POWER AMPLIFIER

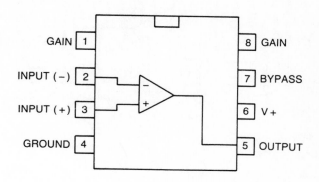

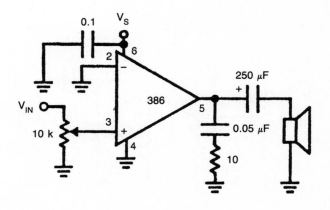

Amplifier with gain = 20 V/V (26 dB)

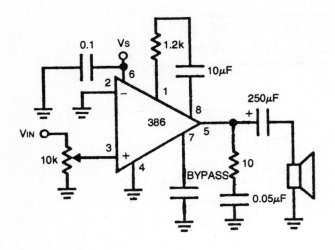

Amplifier with gain = 50 V/V (34 dB)

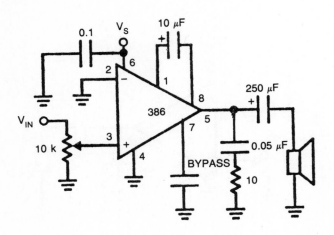

Amplifier with gain = 200 V/V (46 dB)

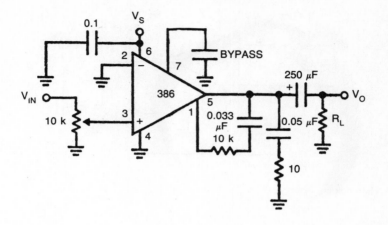

Amplifier with bass boost

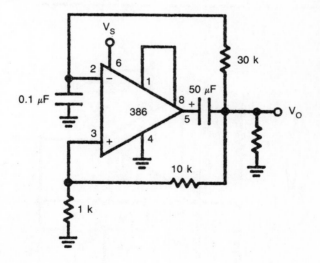

Square-wave oscillator

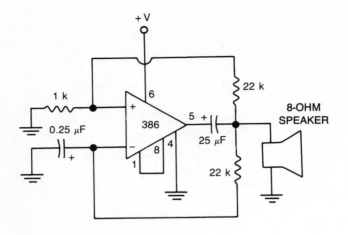

High volume warning siren

387—LOW-NOISE DUAL PREAMPLIFIER

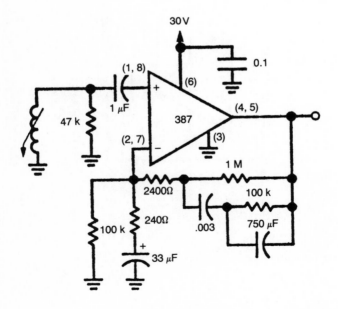

Phono preamp

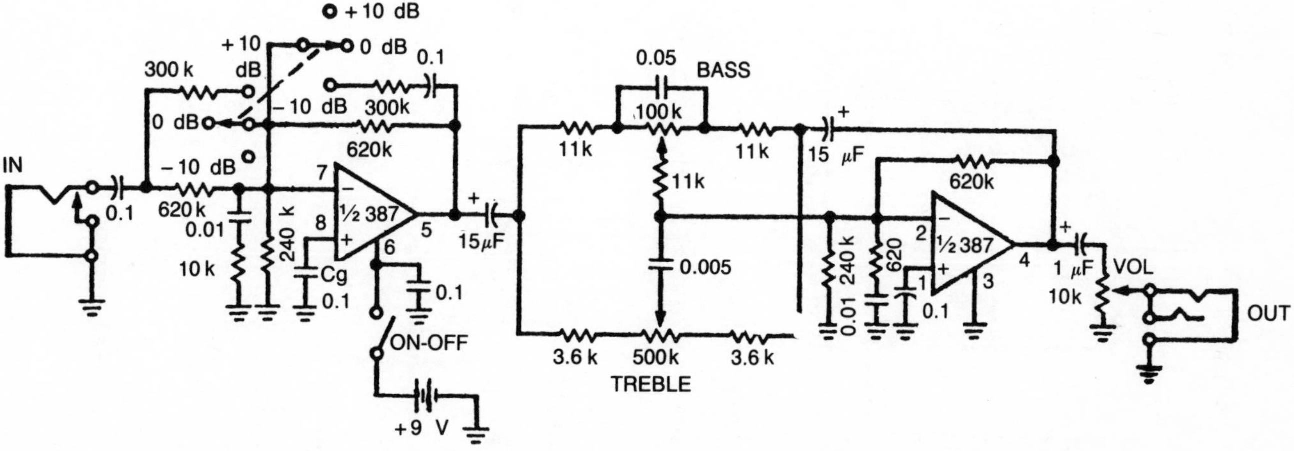

Acoustic pickup preamp

225

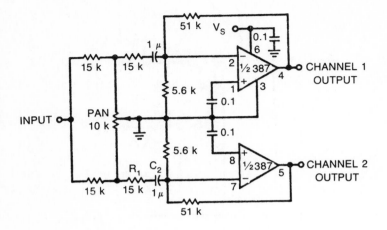

Two-channel panning circuit

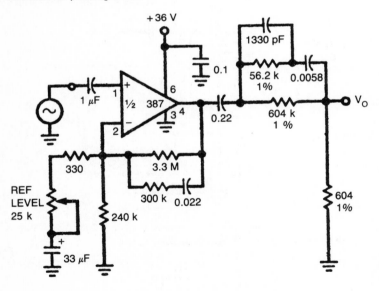

Inverse RIAA response generator

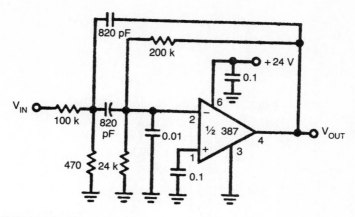

20kHz bandpass active filter

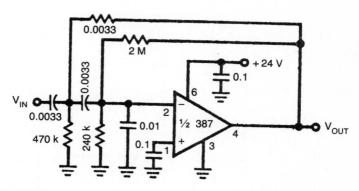

Rumble filter

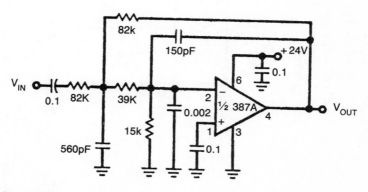

Scratch filter

227

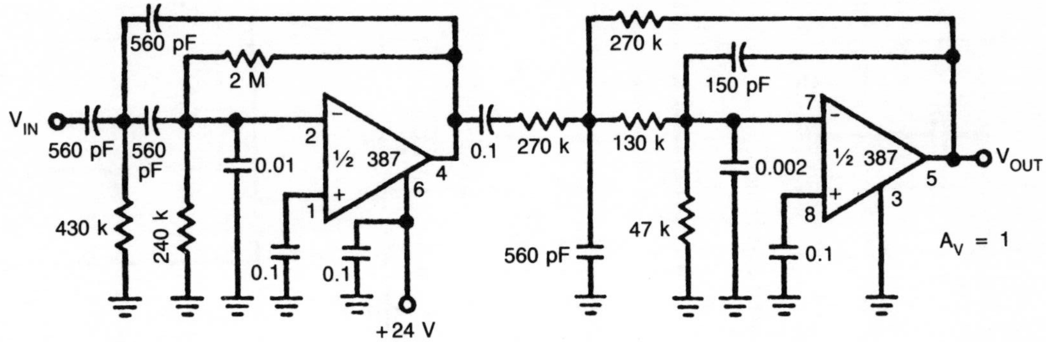

Speech filter (300 Hz - 3 kHz bandpass)

228

387A—LOW-NOISE DUAL PREAMPLIFIER

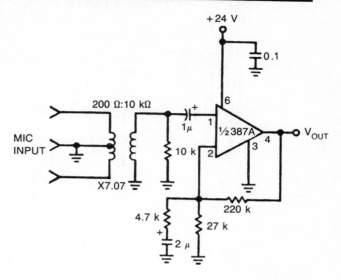

Balanced input mic preamp

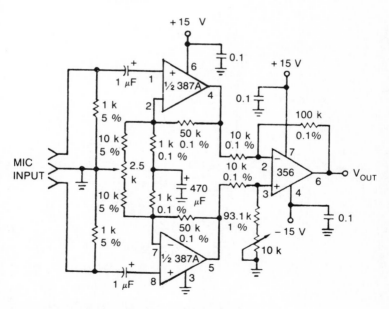

Low-noise transformerless balanced mic preamp

388—1.5-WATT AUDIO POWER AMPLIFIER

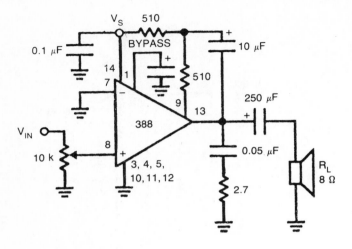

Load returned to ground amplifier with gain = 20

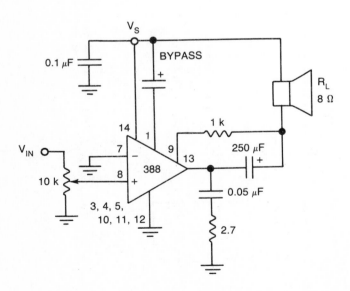

Load returned to Vs amplifier with gain = 20

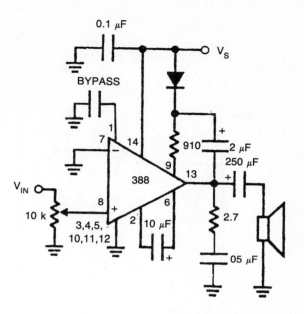

Amplifier with gain = 200

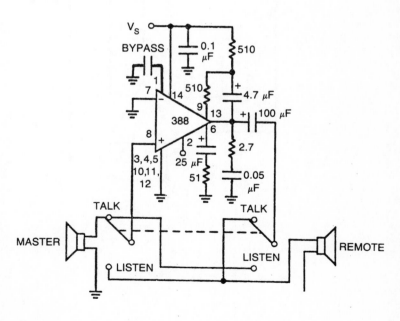

Intercom

Bridge amp

389—LOW-VOLTAGE AUDIO POWER AMPLIFIER WITH NPN TRANSISTOR ARRAY

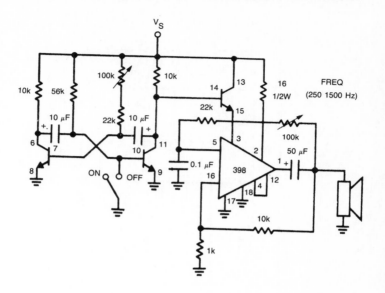

Siren

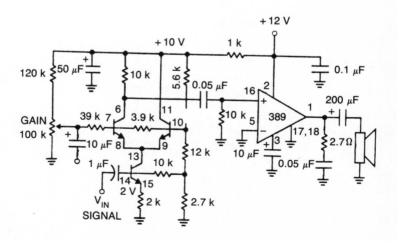

Voltage-controlled amplifier or tremolo circuit

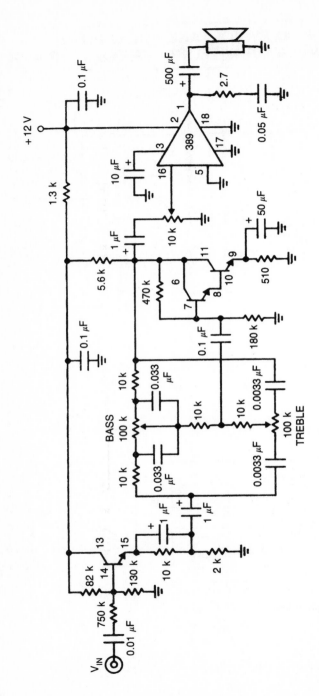

Ceramic phono amplifier with tone controls

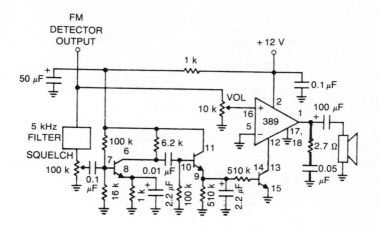

FM scanner noise squelch circuit

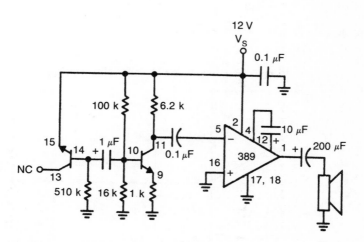

Noise generator using zener diode

390—ONE-WATT LOW-VOLTAGE AUDIO AMPLIFIER

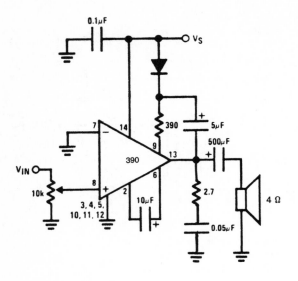

Amplifier with gain = 200 and minimum C_B

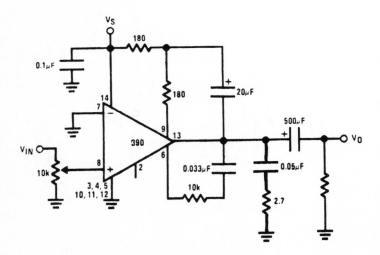

Amplifier with bass boost

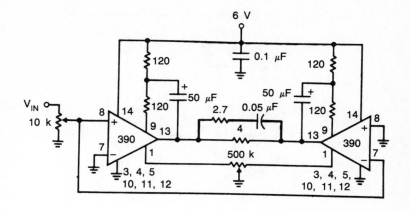

2.5W bridge amplifier

398—MONOLITHIC SAMPLE
AND HOLD CIRCUIT

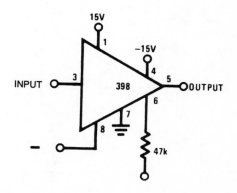

2-channel switch

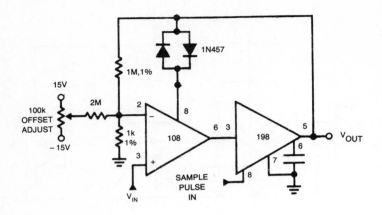

X1000 sample & hold

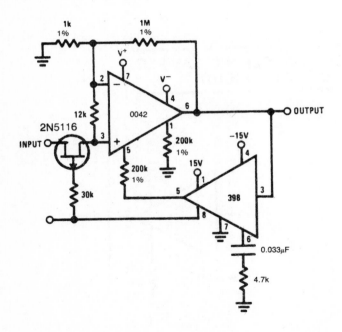

Reset stabilized amplifier (gain of 1000)

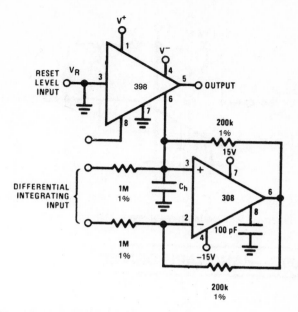

Integrator with programmable reset level

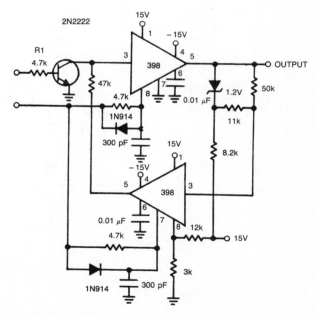

Staircase generator

239

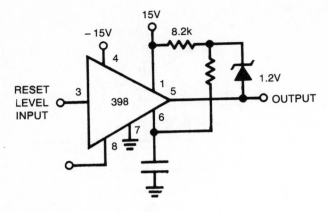

Ramp generator with variable reset level

TL490C—BARGRAPH DRIVER

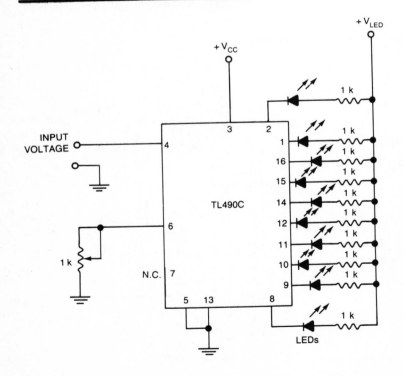

Bargraph

540—INTEGRATED POWER DRIVER

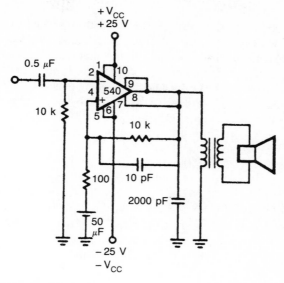

1-Watt power amplifier

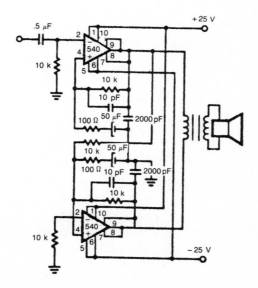

3-Watt power amplifier

555—INTEGRATED TIMER

The 555 timer is a linear device, it's output signal can be used directly by digital circuitry; generally in one of two modes. In the monostable mode, the device functions as a monostable multivibrator, or one-shot timer. The output is normally LOW, but it goes HIGH when the chip is triggered by a negative-going pulse on pin #2. The output remains HIGH for a period determined by an external resistor and capacitor, according to this formula;

$$T = 1.1RC$$

In the astable mode, the 555 functions as an astable multivibrator, or rectangle-wave generator. The output switches back and forth between LOW and HIGH at a regular rate, determined by two external resistors and a capacitor. The HIGH time formula is;

$$Th = 0.693C(Ra + Rb)$$

The LOW time formula is;

$$Tl = 0.693CRa$$

The total cycle time is simply the sum of these two component times;

$$Tt = Th + Tl$$
$$= 0.693C(Ra + 2Rb)$$

The frequency is the reciprocal of the cycle time, so;

$$F = 1/Tt$$
$$= 1/(0.693C(Ra + 2\ Rb))$$

There are a number of variations available, including the 1555 and the 7555 (a CMOS version). The 556 is a single chip containing two 555-type timers. The 558 is a quad timer, containing four slightly simplified 555-type timer stages.

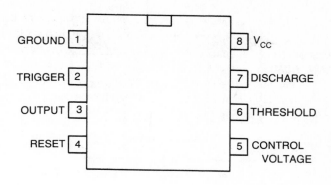

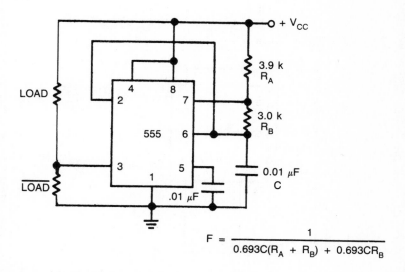

$$F = \frac{1}{0.693C(R_A + R_B) + 0.693CR_B}$$

Astable multivibrator

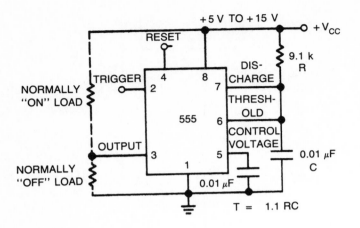

Monostable multivibrator

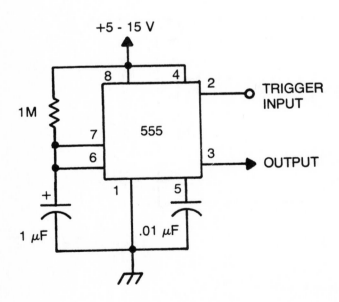

One-shot timer

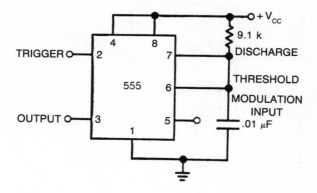

Pulse width modulator

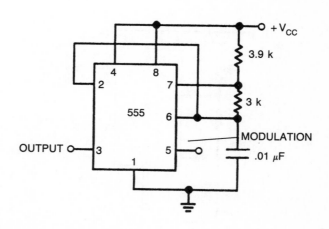

Pulse position modulator

245

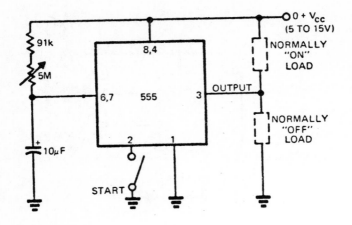

On or off control

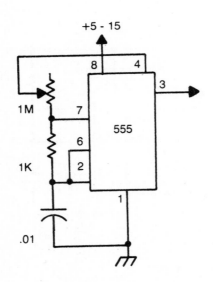

Pulse generator

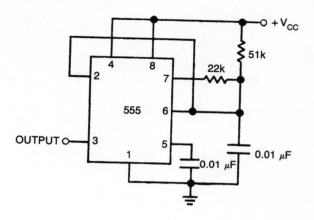

50% duty cycle oscillator

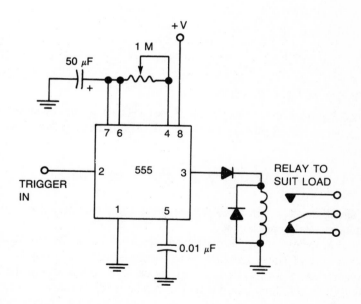

Timed relay

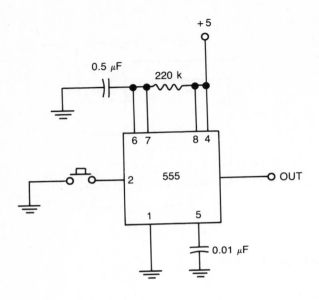

Switch debouncer

TIMED TOUCH SWITCH

The timer (in the monostable mode) will be activated by merely touching a fingertip to the touch pad. The output will go HIGH for a period equal to;

$$T = 1.1RC$$

Using the component values shown here, the timing period will be approximately one second.

For safety's sake, any touch switch circuit should be operated on battery power ONLY. Never use an ac-to-dc power supply for a touch switch. An accidental short circuit could result in a painful shock, or possibly even death. Don't take foolish chances.

This circuit may not work reliably in an area with a strong stray ac field. You could make the circuit a little more reliable by connecting a second touch pad to pin #1 (ground). To trigger the timer, use your fingertip to short across the two touch pads.

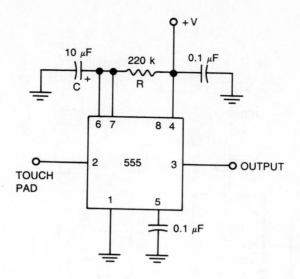

Timed touch switch

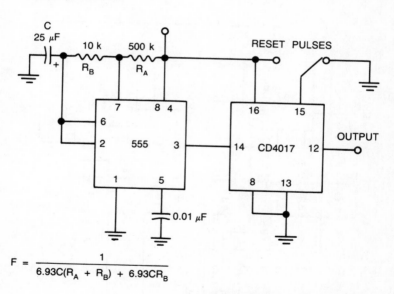

$$F = \frac{1}{6.93C(R_A + R_B) + 6.93CR_B}$$

ADD MORE CD4017 FOR SLOWER
PULSE RATES. EACH CD4017 DIVIDES
THE INPUT FREQUENCY BY TEN.

Very low rate pulse generator

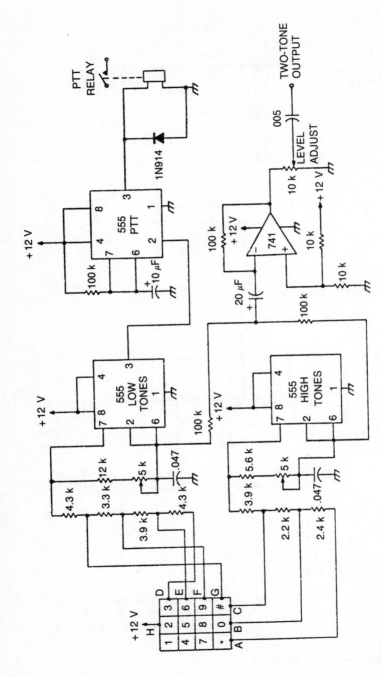

Telephone dialing tone encoder

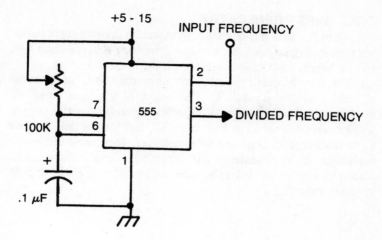

Frequency divider

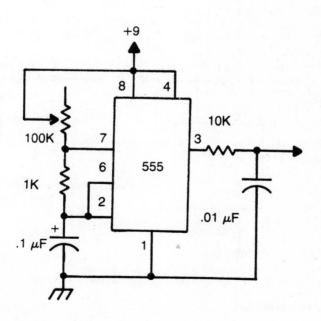

Triangle-wave generator

TONE BURST GENERATOR

When the switch is momentarily closed, a brief string of pulses will appear at the output. If the output drives a speaker, a short tone will be heard, each time the switch is depressed.

The tone frequency is determined by the values of capacitor C_a, and resistors R_a and R_b. The length of the tone burst will be determined by the value of capacitor C_b and the setting of the 250 k potentiometer. The tone burst will last for the time it takes C_b to discharge through the potentiometer. Increasing either the resistance or the capacitance will increase the length of the burst. Similarly, shorter bursts can be achieved by reducing either or both of these values.

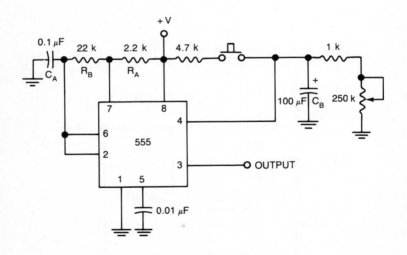

Tone burst generator

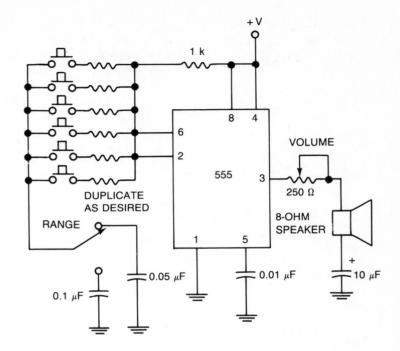

Toy organ

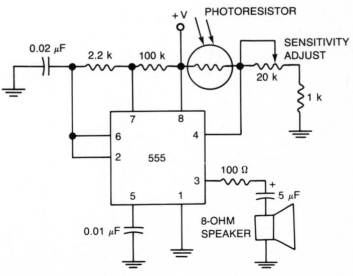

Light on alarm

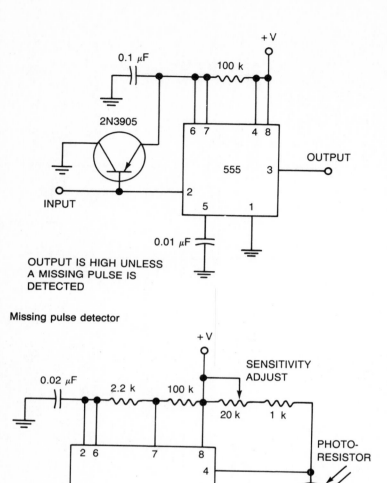

OUTPUT IS HIGH UNLESS
A MISSING PULSE IS
DETECTED

Missing pulse detector

Light off alarm

254

556—DUAL 555-TYPE TIMER

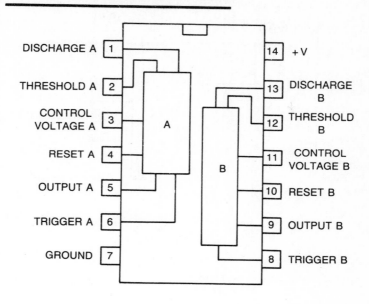

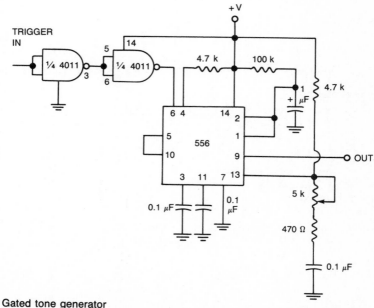

Gated tone generator

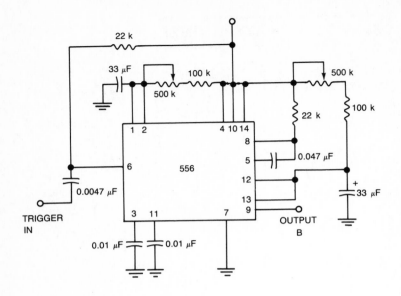

Extended range monostable timer

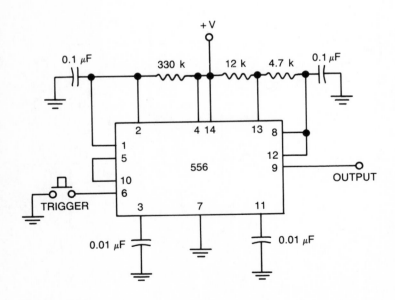

Tone burst generator

256

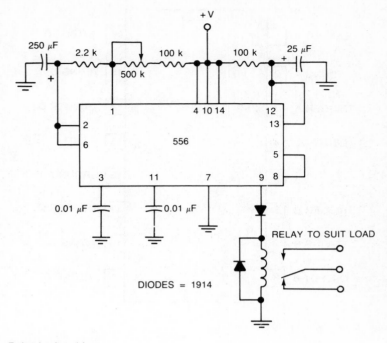

Pulsed relay driver

558 QUAD TIMER

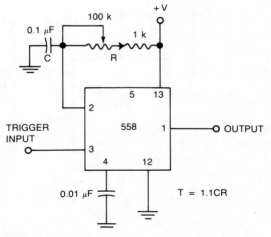

Monostable multivibrator

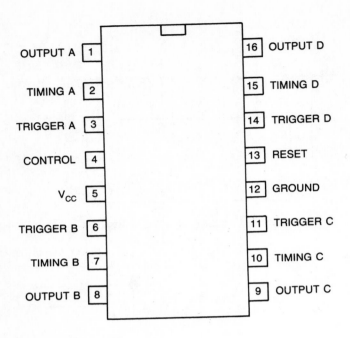

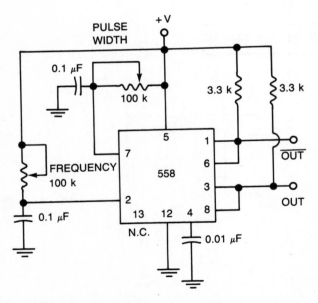

Rectangle-wave generator with independent pulse width control

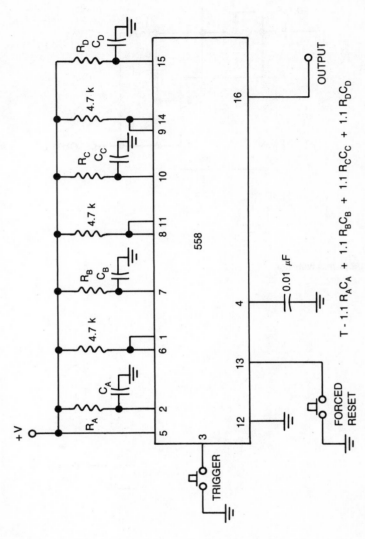

$$T - 1.1\, R_A C_A\ +\ 1.1\, R_B C_B\ +\ 1.1\, R_C C_C\ +\ 1.1\, R_D C_D$$

Extender range timer

259

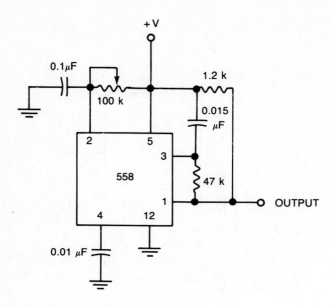

Astable multivibrator

260

566—VCO (VOLTAGE CONTROLLED OSCILLATOR)

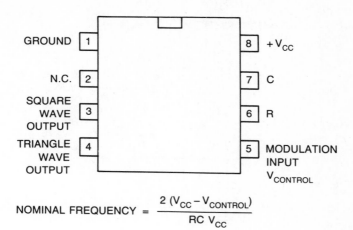

NOMINAL FREQUENCY = $\dfrac{2\,(V_{CC} - V_{CONTROL})}{RC\,V_{CC}}$

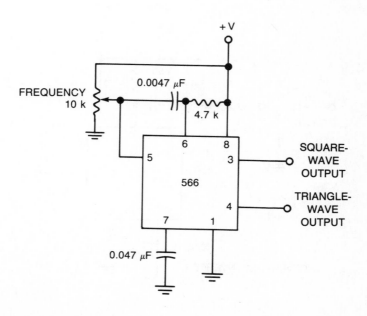

Two waveform oscillator

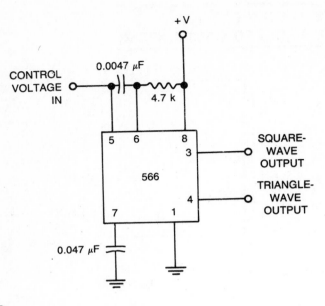

VCO

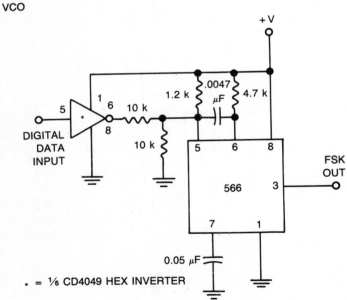

* = ⅙ CD4049 HEX INVERTER

(Frequency Shift Keying for encoding digital data in audible tones.)

262

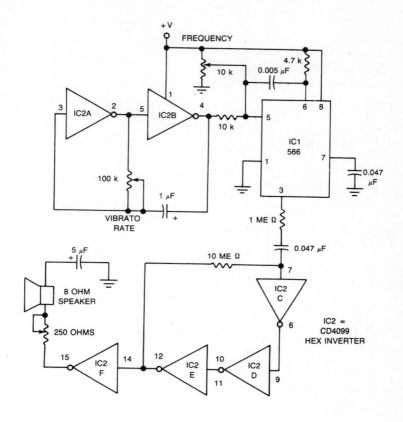

Vibrato tone generator

THE 567 TONE DECODER

This device contains a PLL, or Phase-Locked Loop. The output goes LOW when the input frequency is equal to (or very close to) the center frequency (F_0) set by the timing resistor (R — pin #5) and the timing capacitor (C — pin #6). The formula is:

$$F_0 = 1.1/RC$$

The center frequency may be set anywhere from 0.01 Hz up to 500 kHz. For reliable operation the timing resistor's value should be kept between 2k and 20k.

567—TONE DECODER

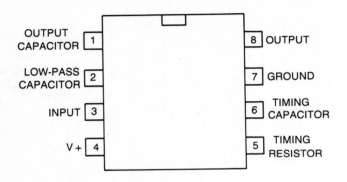

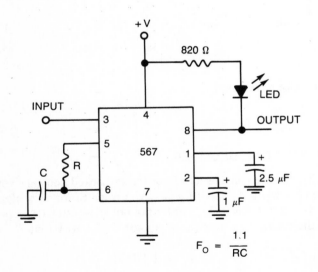

$$F_O = \frac{1.1}{RC}$$

Tone decoder

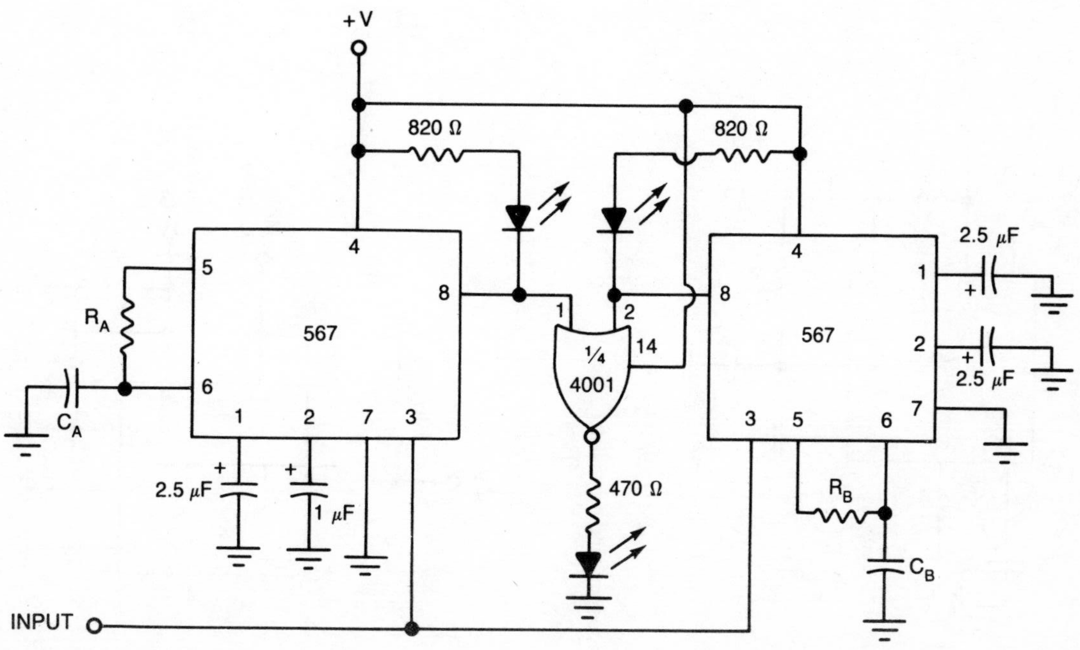

Narrow range tone decoder

265

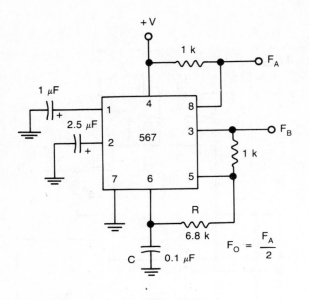

Dual frequency output oscillator

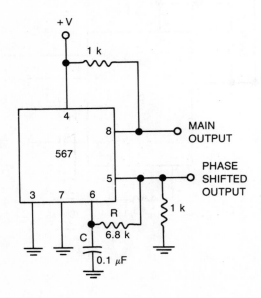

Oscillator with phase shifted output

760—HIGH SPEED
DIFFERENTIAL COMPARATOR

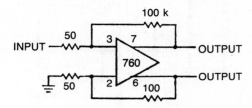

Level detector with hysteresis

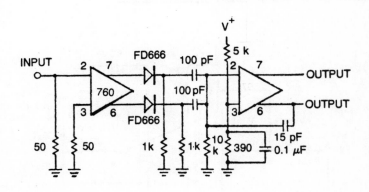

Zero crossing detector

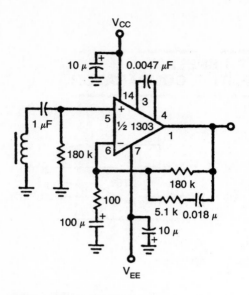

Tape preamp (NAB, 1-7/8 & 3-3/4 IPS)

1306—HALF WATT
INTEGRATED AUDIO AMPLIFIER

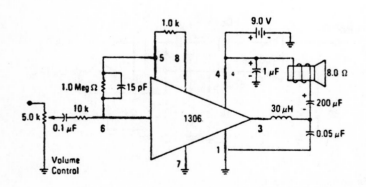

AM-FM radio, audio section

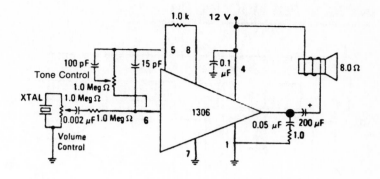

Phonograph amplifier (ceramic cartridge)

1422—MONOLITHIC TIMER WITH EXTERNALLY ADJUSTABLE THRESHOLD

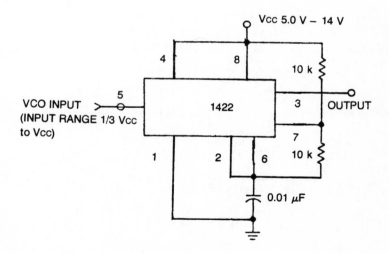

Voltage controlled oscillator

1494—LINEAR FOUR-QUADRANT INTEGRATED MULTIPLIER

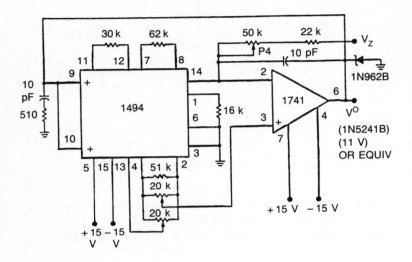

Square root circuit

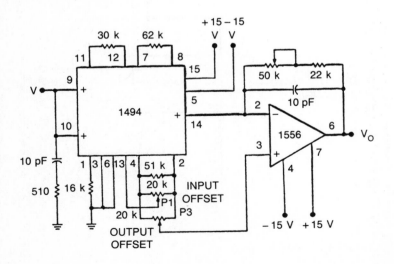

Squaring circuit

1495—WIDEBAND LINEAR FOUR-QUADRANT INTEGRATED MULTIPLIER

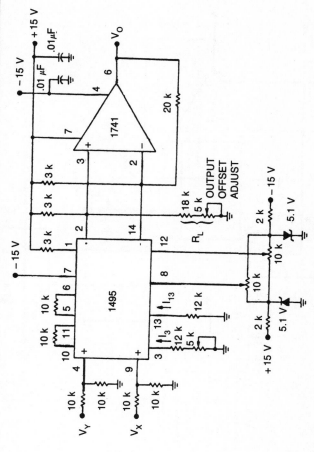

Multiplier with op-amp level shift

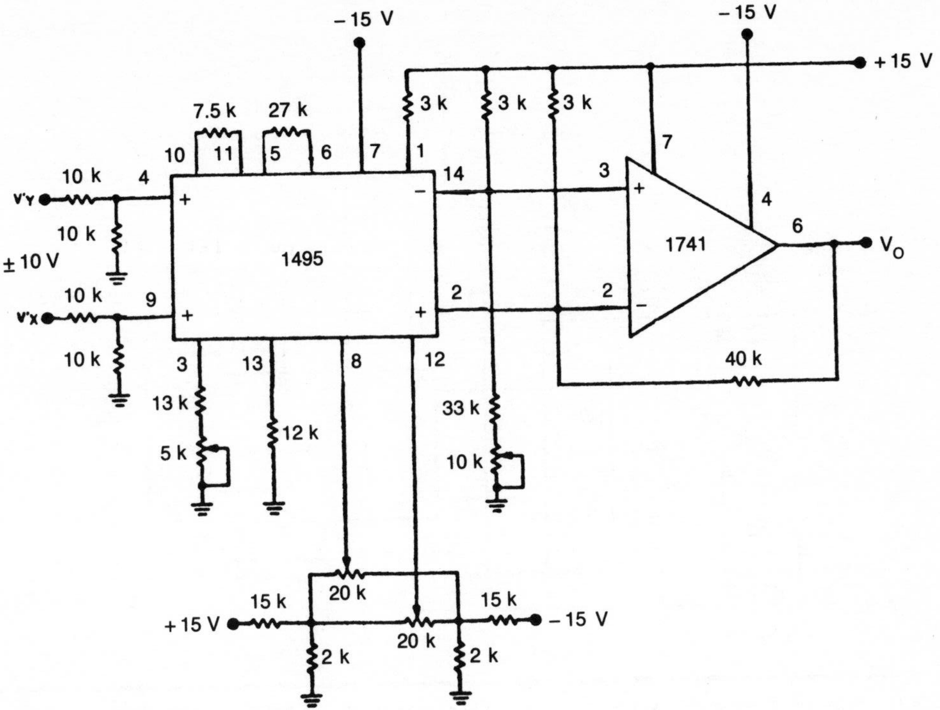

Multiplier with improved linearity

272

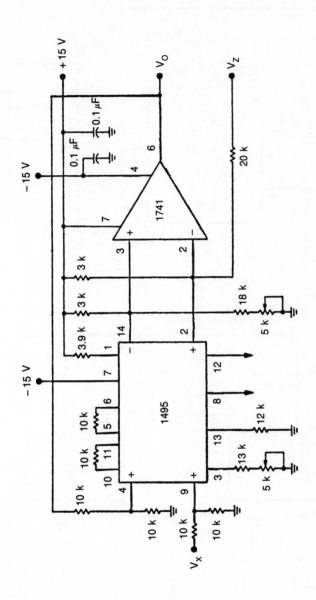

Divide circuit

273

1555—INTEGRATED TIMER

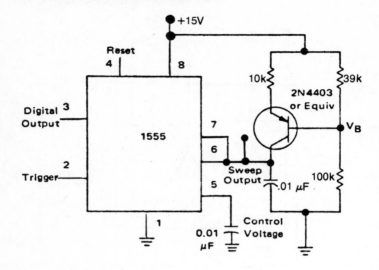

Linear voltage sweep circuit

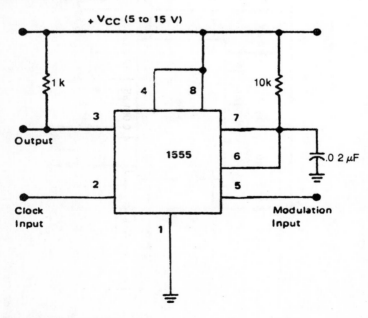

Pulse width modulator

274

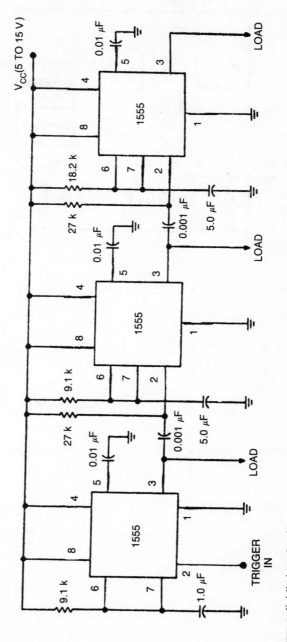

Sequential timing circuit

275

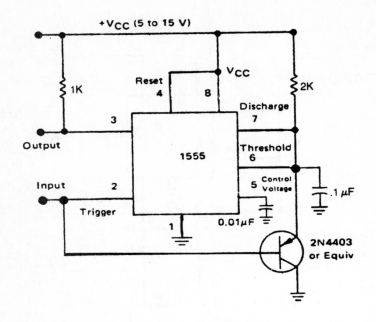

Missing pulse detector

1877—DUAL CHANNEL POWER AUDIO AMPLIFIER

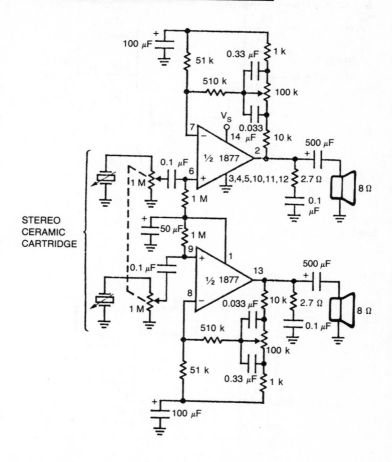

Stereo phonograph amplifier with bass tone control

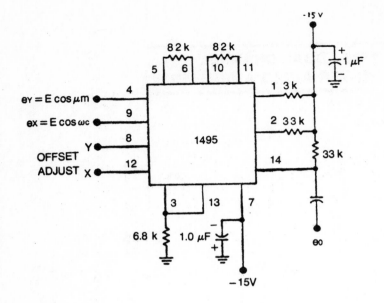

Balanced modulator

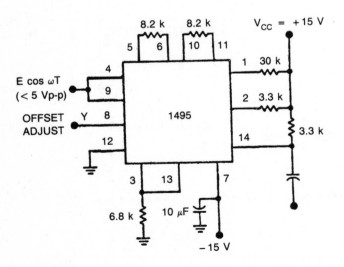

Frequency doubler

278

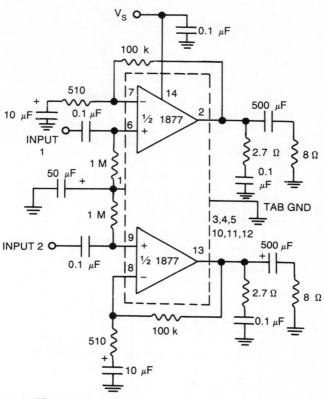

Stereo amplifier with $A_V = 200$

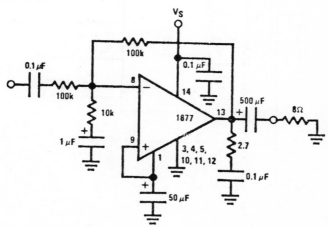

Inverting unity gain amplifier

279

Square root circuit

XR-2206—FUNCTION GENERATOR

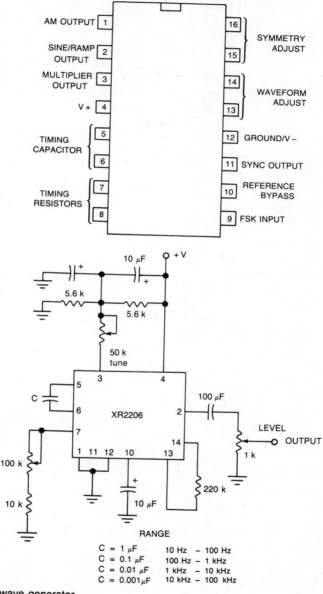

Sine wave generator

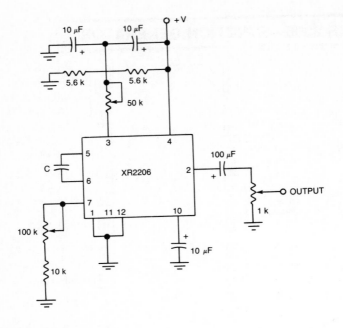

Triangle-wave generator

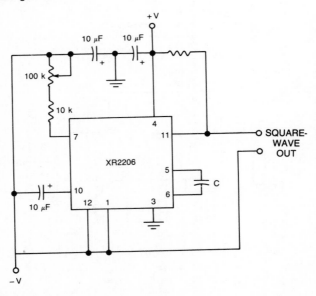

Square-wave generator

XR-2208—OPERATIONAL MULTIPLIER

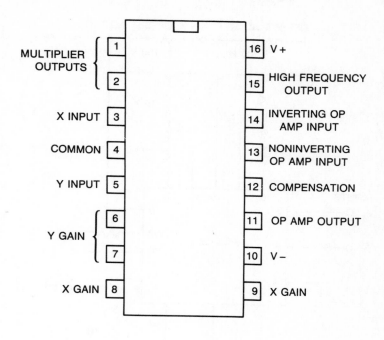

MULTIPLIER OUTPUTS
- 1
- 2

X INPUT — 3

COMMON — 4

Y INPUT — 5

Y GAIN
- 6
- 7

X GAIN — 8

16 — V +

15 — HIGH FREQUENCY OUTPUT

14 — INVERTING OP AMP INPUT

13 — NONINVERTING OP AMP INPUT

12 — COMPENSATION

11 — OP AMP OUTPUT

10 — V −

9 — X GAIN

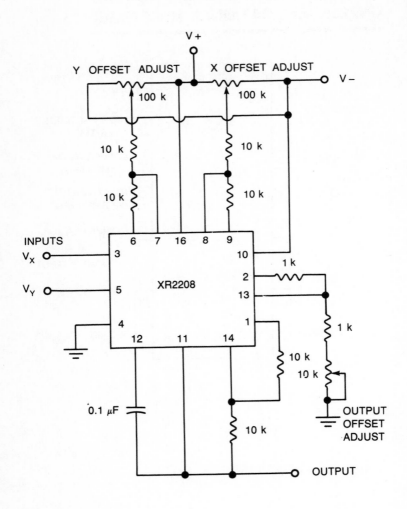

Multiplier

284

XR2240—PROGRAMMABLE TIMER

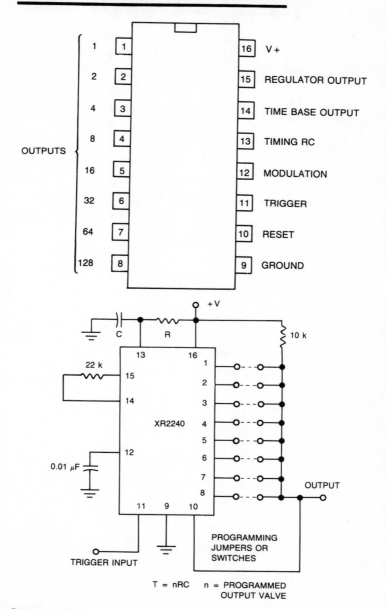

Programmable monostable

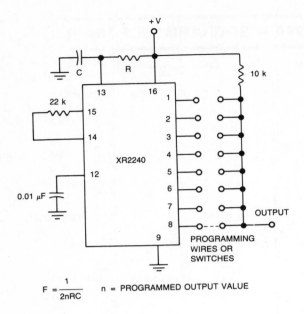

$$F = \frac{1}{2nRC} \qquad n = \text{PROGRAMMED OUTPUT VALUE}$$

Programmable astable multivibrator

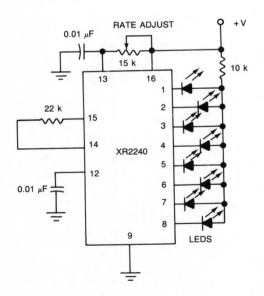

Multiple LED flasher

HA-2400—PROGRAMMABLE AMPLIFIER

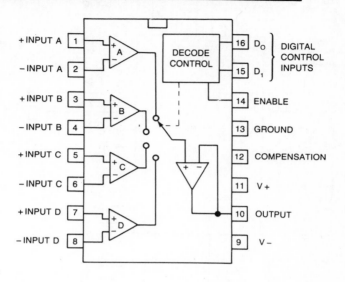

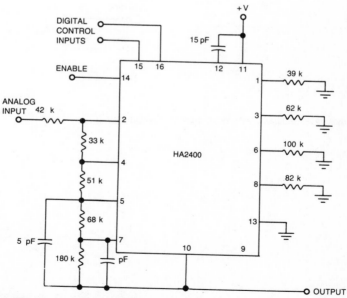

Programmable inverting amplifier

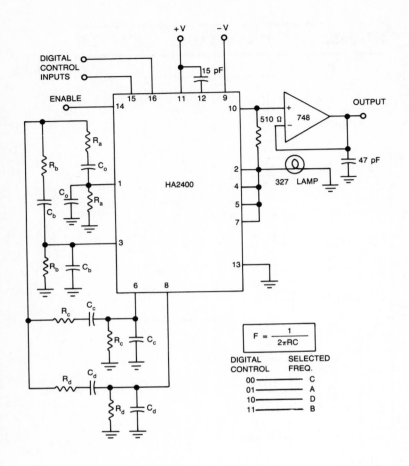

Sine wave oscillator with programmable frequency

3380—EMITTER COUPLED INTEGRATED ASTABLE MULTIVIBRATOR

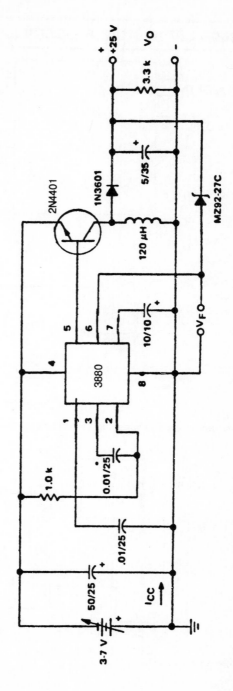

Typical application in 3 - 25 V dc-dc converter configuration

LM3909—LED FLASHER/OSSCILLATOR

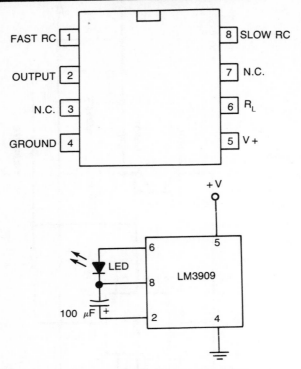

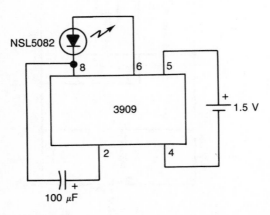

Decrease capacitance to speed up flash rate LED flasher

Minimum power at 1.5V

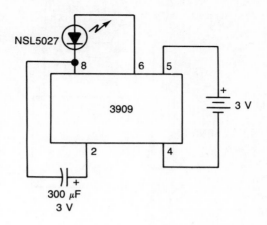

3V flasher

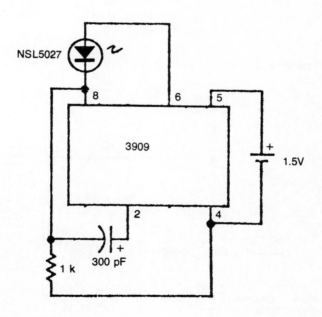

Fast blinker

291

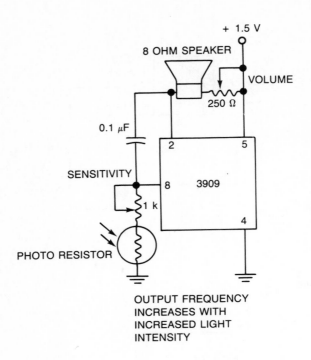

Photo sensitive oscillator

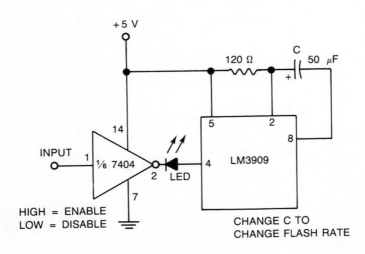

TTL controlled LED flasher

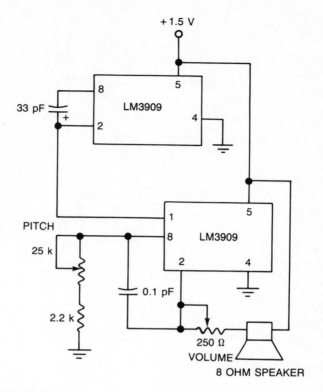

Bird chirp simulator

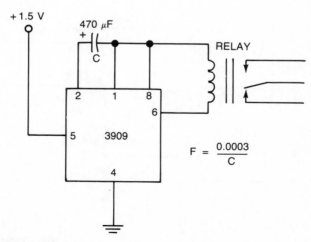

$$F = \frac{0.0003}{C}$$

Relay driver oscillator

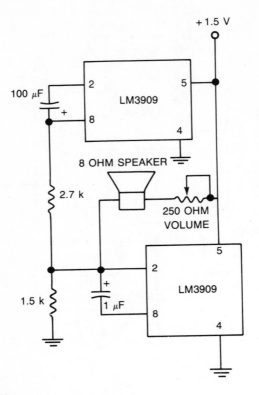

Two tone siren

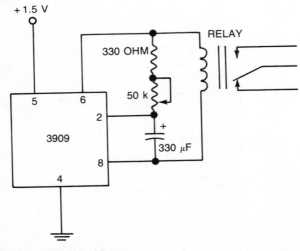

Variable frequency relay driver

294

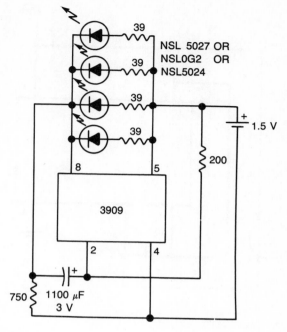

4 parallel LEDs

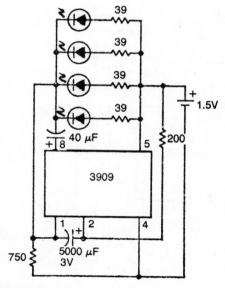

High efficiency parallel circuit

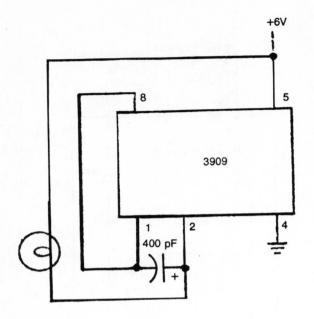

Incandescent bulb flasher

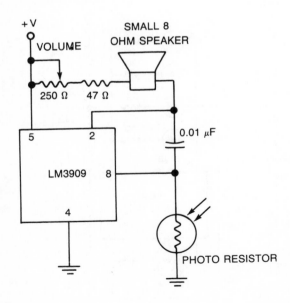

Light activated oscillator

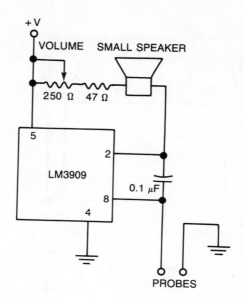

Liquid sensor/alarm

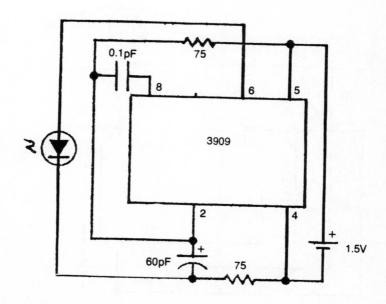

LED booster

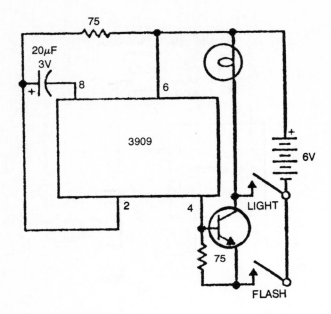

Emergency lantern/flasher

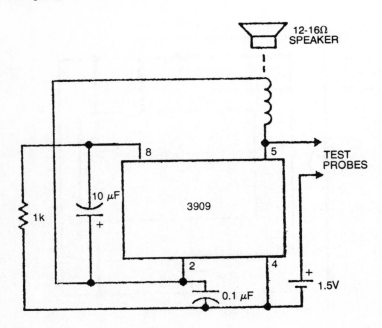

"Buzz box" continuity and coil checker

298

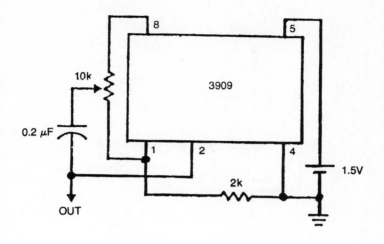

1 kHz square-wave oscillator

MM5837N—NOISE GENERATOR

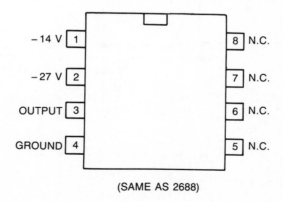

(SAME AS 2688)

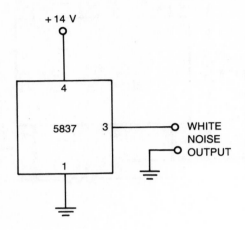

White noise generator

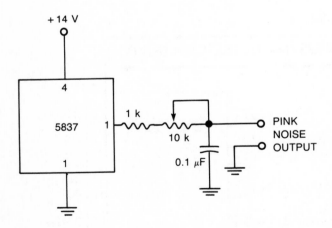

Pink noise generator

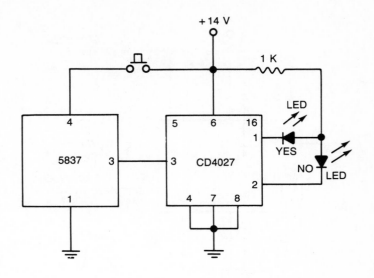

Random "decision maker"

8038—FUNCTION GENERATOR

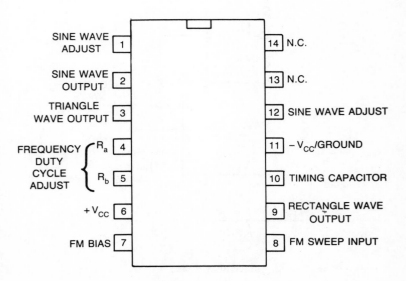

SINE WAVE ADJUST	1	14 N.C.
SINE WAVE OUTPUT	2	13 N.C.
TRIANGLE WAVE OUTPUT	3	12 SINE WAVE ADJUST
FREQUENCY DUTY CYCLE ADJUST R_a	4	11 $-V_{CC}$/GROUND
R_b	5	10 TIMING CAPACITOR
$+V_{CC}$	6	9 RECTANGLE WAVE OUTPUT
FM BIAS	7	8 FM SWEEP INPUT

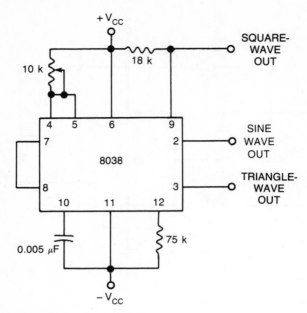

Simple function generator

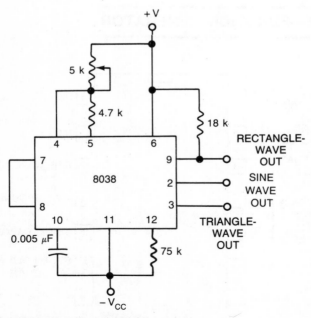

Function generator with variable duty cycle

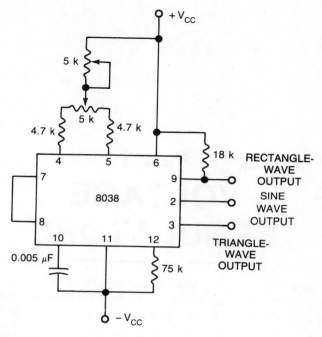

Function generator with variable frequency and duty cycle

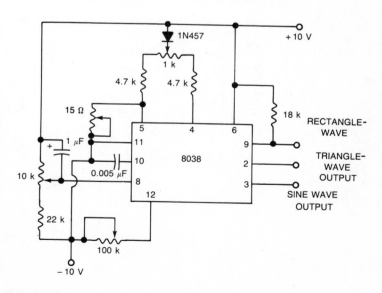

Audio oscillator

303

Part 3
VOLTAGE REGULATORS

It wasn't too long ago when voltage regulator circuits involved several transistors (or tubes!), a few diodes, resistors, capacitors and a lot of space and cost. Now a single chip replaces all that.

109—5 VOLT
INTEGRATED VOLTAGE REGULATOR

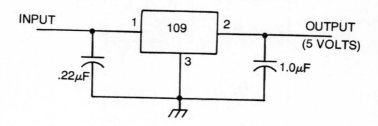

Fixed 5 volt regulator

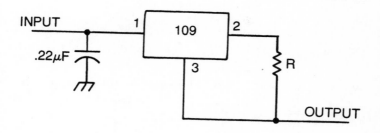

Current regulator (output current varies with value of resistor R)

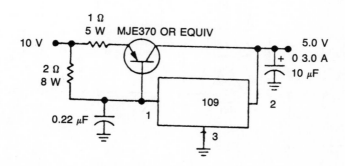

5.0-volt, 3.0-ampere regulator (with plastic boost transistor)

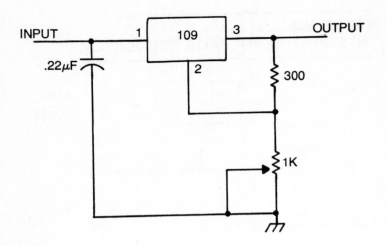

Regulator with adjustable output

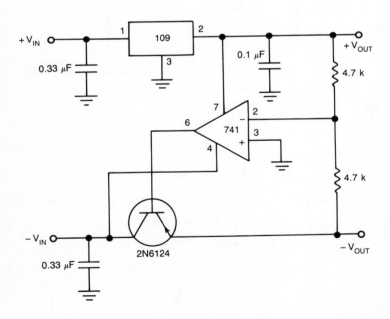

Tracking voltage regulator

306

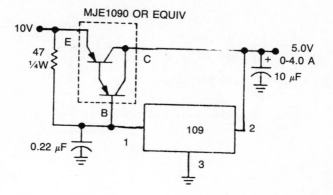

5.0 volt, 4.0-ampere transistor (with plastic Darlington boost transistor)

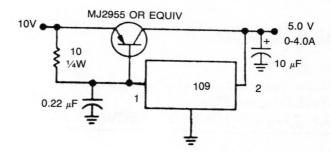

5.0-volt, 10-ampere regulator

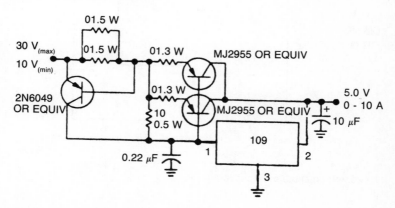

5.0-volt, 10-ampere regulator (with short-circuit current limiting for safe-area protection of pass transistors)

307

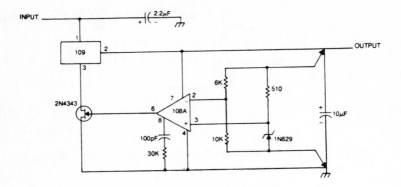

High stability (0.01%) regulator

117—INTEGRATED ADJUSTABLE VOLTAGE REGULATOR

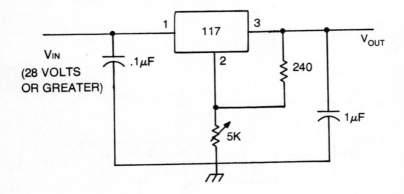

1.2-25 volt adjustable output regulator

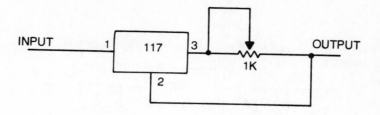

Precision current limiter

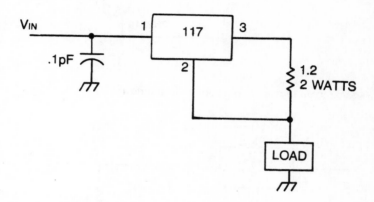

One amp current regulator

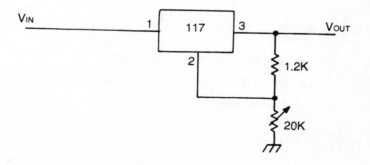

1.2-20 volt adjustable regulator with 4 mA minimum load current

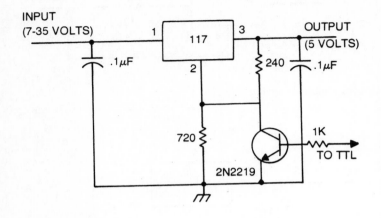

Five volt logic regulator with electronic shutdown

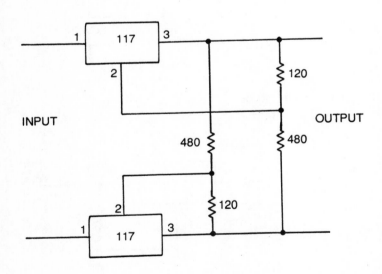

AC voltage regulator

310

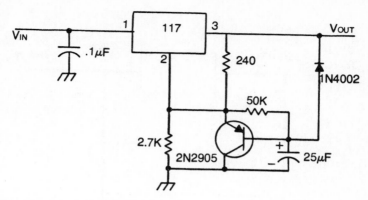

Slow turn on 15 volt regulator

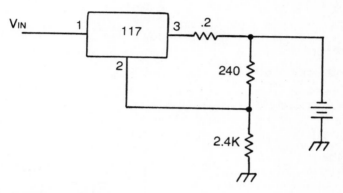

12 volt battery charger

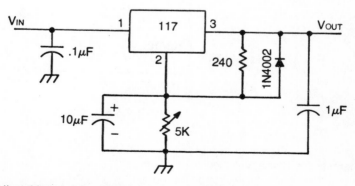

Adjustable regulator with improved ripple suppression

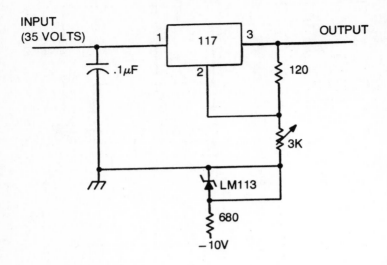

0 to 30 volt regulator

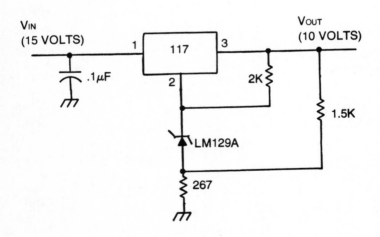

Highly stable 10 volt regulator

123—5 VOLT 3 AMP
POSITIVE INTEGRATED REGULATOR

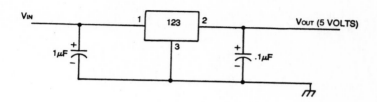

Basic 3 amp regulator

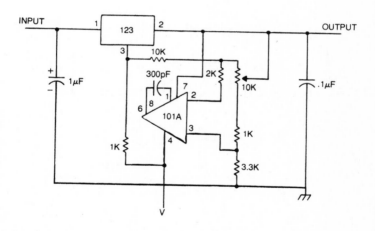

0-10 volt adjustable regulator

LM317—1.2-37 VOLT REGULATOR

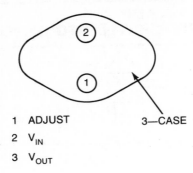

1 ADJUST

2 V_{IN}

3 V_{OUT}

3—CASE

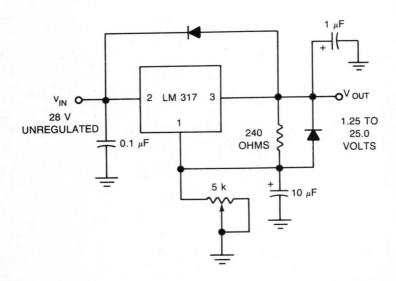

Variable voltage regulator

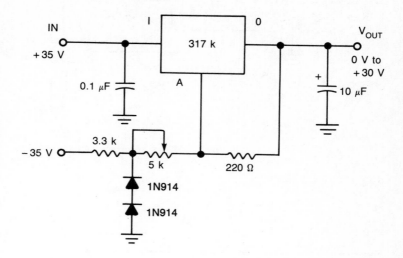

Variable voltage regulator

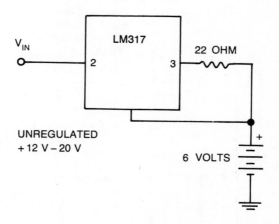

NICAD battery charger

315

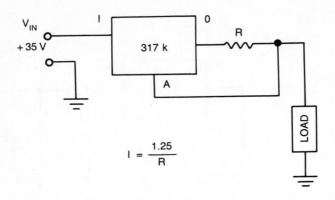

$$I = \frac{1.25}{R}$$

Precision current limiter

337T—1.2 TO —37 VOLT REGULATOR

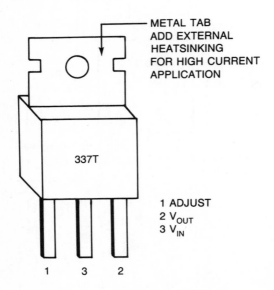

METAL TAB
ADD EXTERNAL
HEATSINKING
FOR HIGH CURRENT
APPLICATION

337T

1 ADJUST
2 V_{OUT}
3 V_{IN}

1   3   2

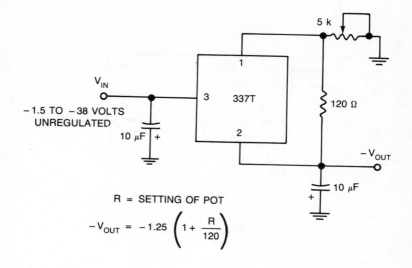

R = SETTING OF POT

$$-V_{OUT} = -1.25 \left(1 + \frac{R}{120}\right)$$

Adjustable negative voltage regulator

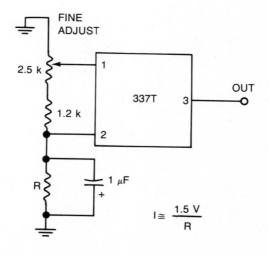

$$I \cong \frac{1.5 \text{ V}}{R}$$

Constant current source

340—SERIES VOLTAGE REGULATOR

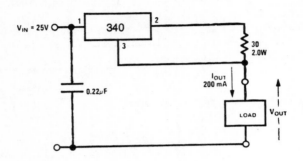

Current source

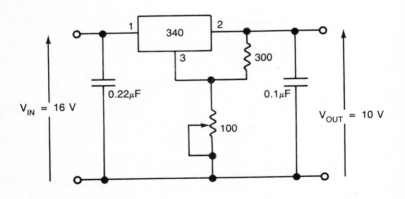

10 V regulator

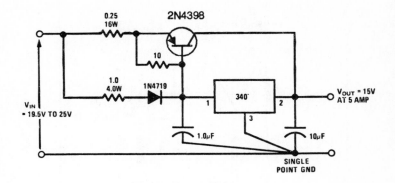

15 V 5.0 A regulator with short circuit current limit

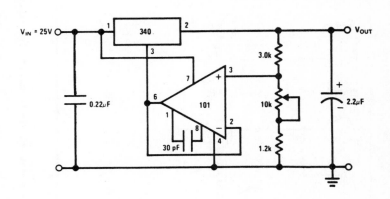

Variable output regulator

319

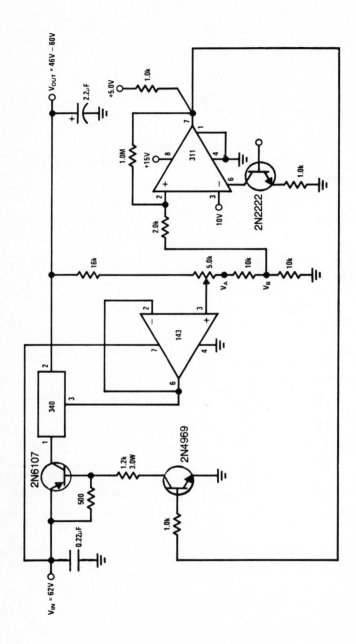

Variable high voltage regulator with short-circuit and overvoltage protection

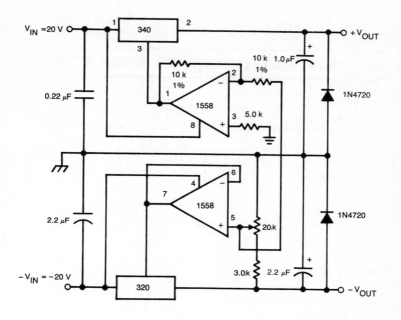

Tracking dual supply ± 5.0 V — ± 18 V

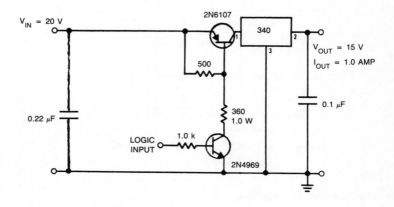

Electronic shutdown circuit

342—INTEGRATED VOLTAGE REGULATOR

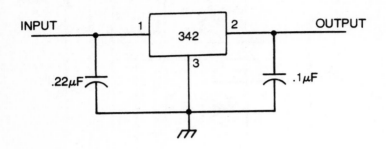

Fixed output regulator

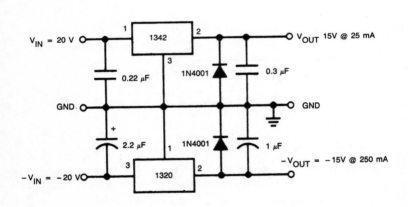

±15V, 250 mA dual power supply

322

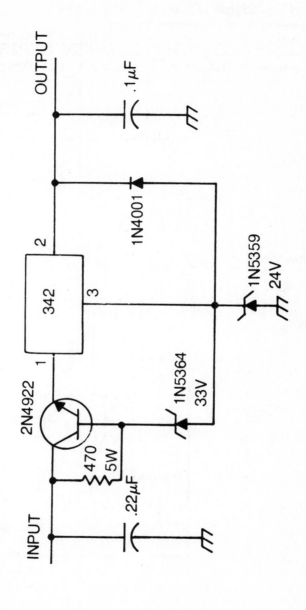

High output regulator

TL431—ADJUSTABLE
SHUNT (ZENER) REGULATOR

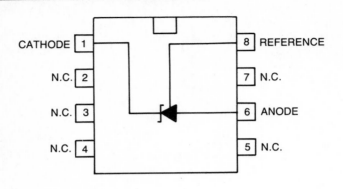

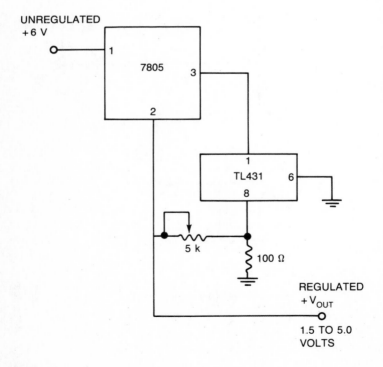

Variable voltage regulator

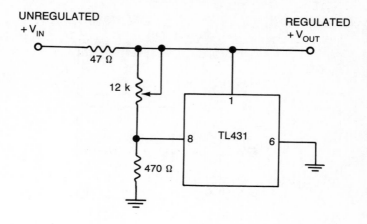

Variable voltage regulator

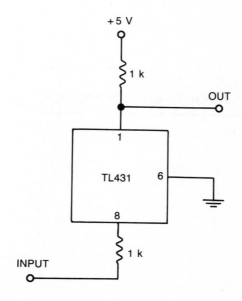

TTL logic level detector

325

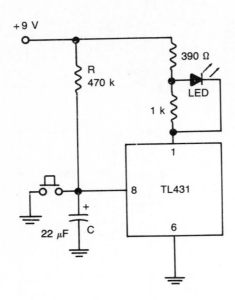

Timer

0070—PRECISION BCD
BUFFERED VOLTAGE REFERENCE

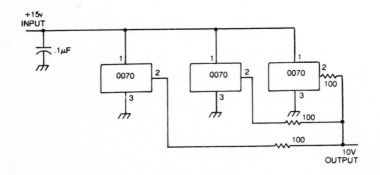

Statistical voltage standard

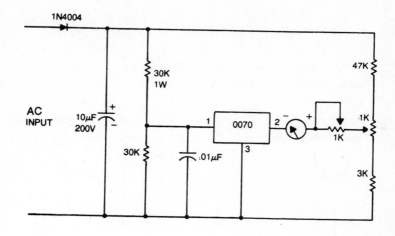

AC voltmeter

7805—+5 VOLT REGULATOR

EXTERNAL HEATSINKING
MAY BE REQUIRED

METAL TAB

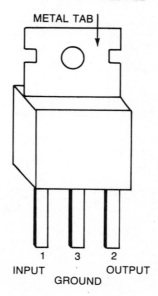

1   3   2
INPUT   OUTPUT
     GROUND

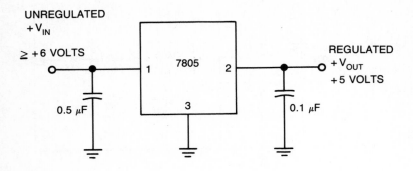

UNREGULATED
+ V_{IN}

≥ +6 VOLTS

7805

0.5 μF

REGULATED
+ V_{OUT}
+5 VOLTS

0.1 μF

+5 Volt regulator

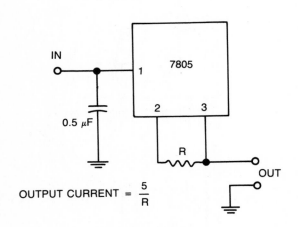

IN

7805

0.5 μF

R

OUT

OUTPUT CURRENT = $\dfrac{5}{R}$

Current regulator

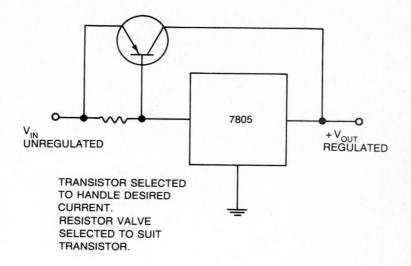

Increased current voltage regulator

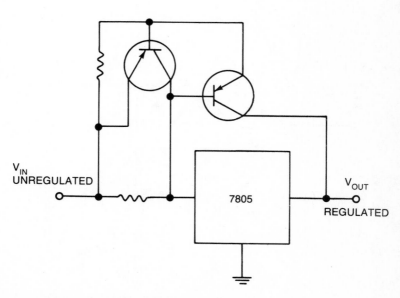

Very high current voltage regulator with short circuit protection

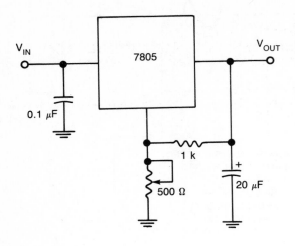

Variable output voltage regulator

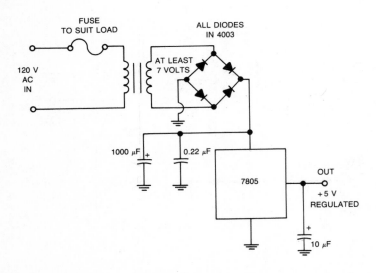

5 Volt regulated power supply

78L05—FIVE VOLT
INTEGRATED VOLTAGE REGULATOR

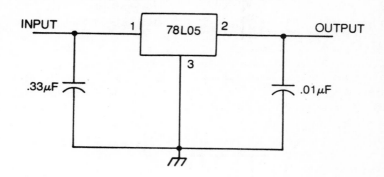

Fixed five volt output regulator

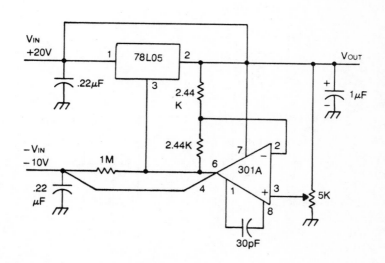

Variable output regulator

7812— +12 VOLT REGULATOR

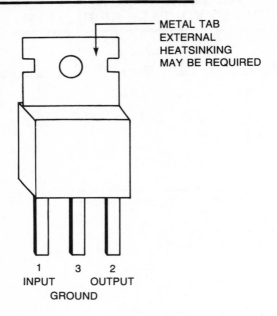

METAL TAB
EXTERNAL
HEATSINKING
MAY BE REQUIRED

1
INPUT

3
GROUND

2
OUTPUT

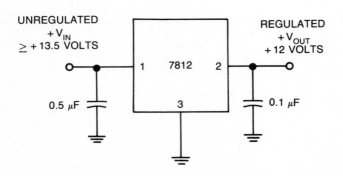

UNREGULATED
+ V_IN
≥ + 13.5 VOLTS

REGULATED
+ V_OUT
+ 12 VOLTS

1   7812   2

3

0.5 μF

0.1 μF

+ 12 Volt regulator

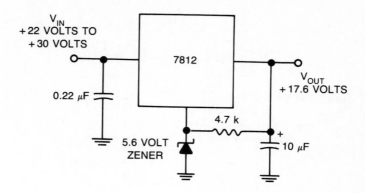

Increased voltage output

7815— +15 VOLT REGULATOR

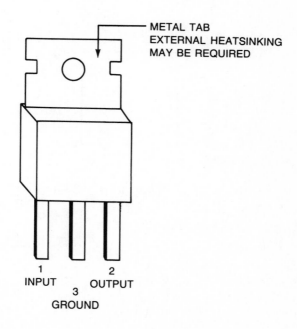

METAL TAB
EXTERNAL HEATSINKING
MAY BE REQUIRED

1
INPUT

2
OUTPUT

3
GROUND

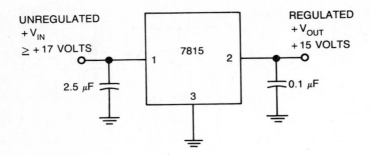

+15 volt regulator

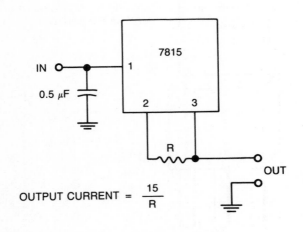

$$\text{OUTPUT CURRENT} = \frac{15}{R}$$

Current regulator

7905—NEGATIVE FIVE VOLT SERIES VOLTAGE REGULATOR

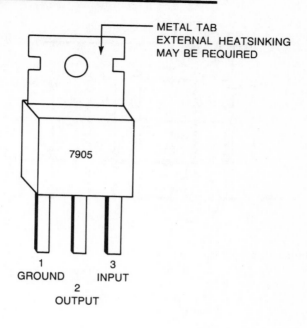

METAL TAB
EXTERNAL HEATSINKING
MAY BE REQUIRED

7905

1
GROUND

3
INPUT

2
OUTPUT

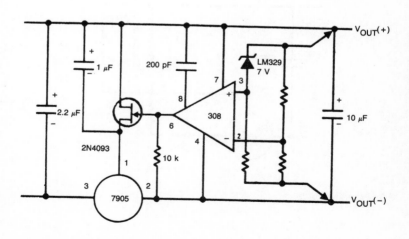

High stability 1 amp regulator

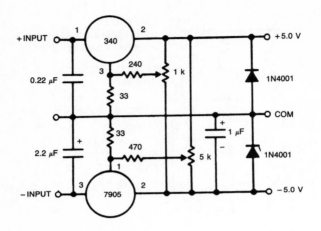

Dual trimmed supply

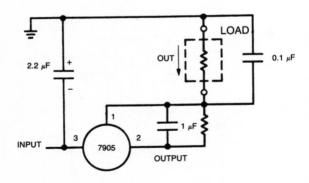

Current source

336

Part 4
CMOS
Integrated Circuits

CMOS digital ICs are the well-known ''4000'' series of devices. They contain more functions per chip than comparable TTL or LS ICs. Most can be used with a power supply ranging from positive 3 to 15 volts. Their only major drawback is their susceptibility to damage from static discharge.

4001—CMOS QUAD NOR GATE

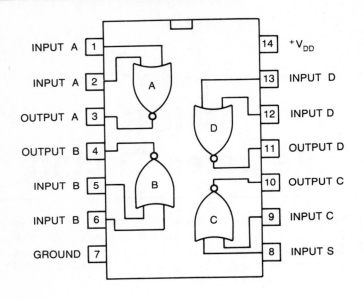

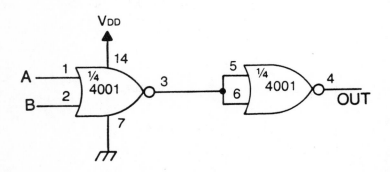

OR gate

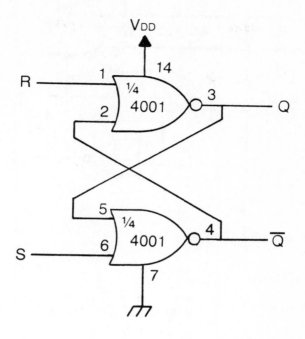

RS latch

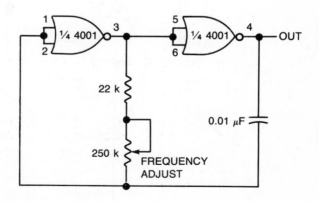

Square-wave generator

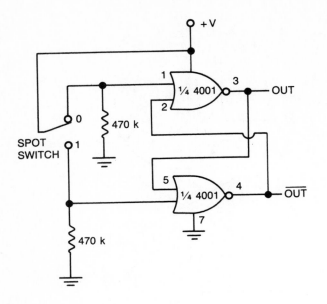

Switch debouncer

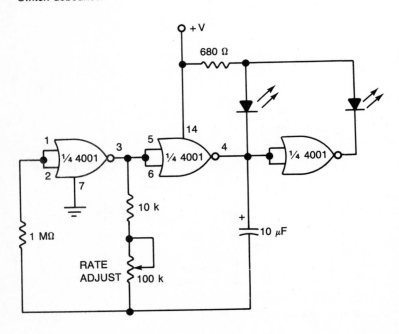

Double LED flasher

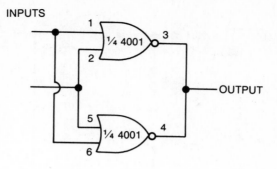

Increased fan-out NOR gate

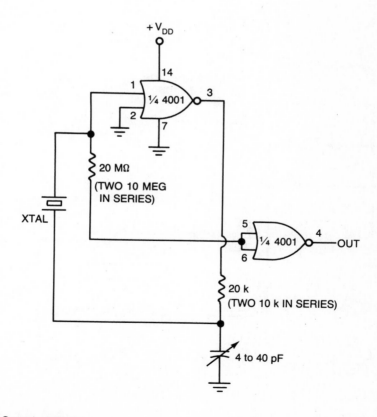

Crystal oscillator

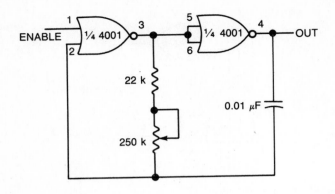

Grated square-wave generator

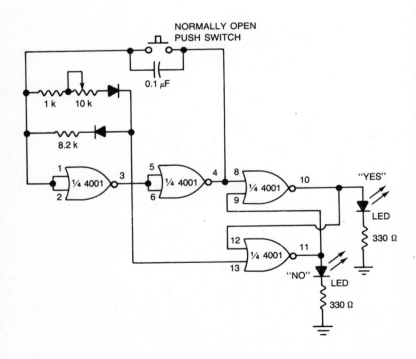

"Decision" maker

342

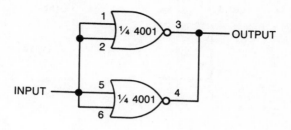

Increased fan-out inverter

4011—CMOS QUAD NAND GATE

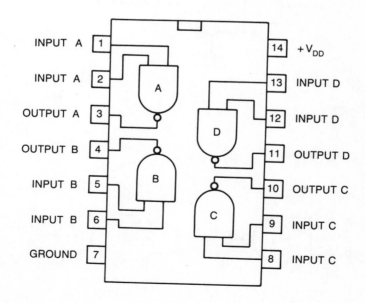

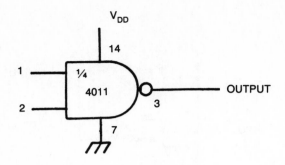

Basic NAND gate

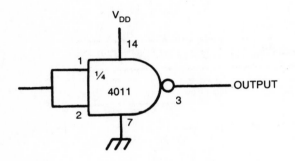

Inverter

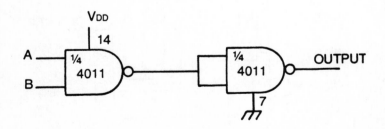

AND gate

344

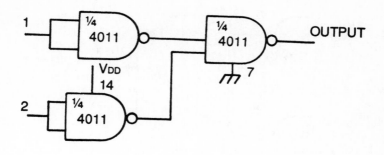

OR gate

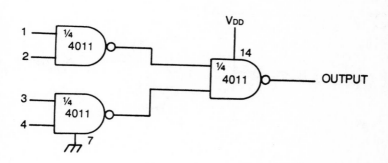

AND/OR gate

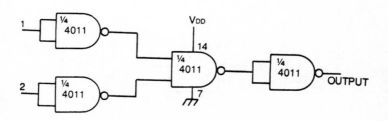

NOR gate

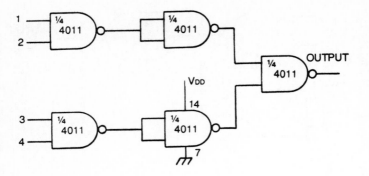

Quad input NAND gate

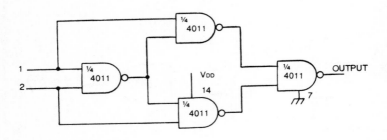

Exclusive OR gate

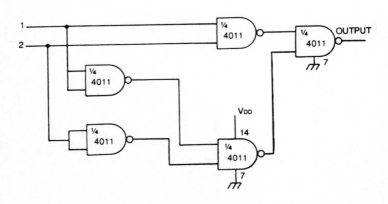

Exclusive NOR gate

346

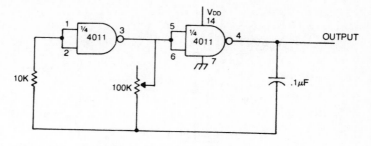

Clock pulse generator

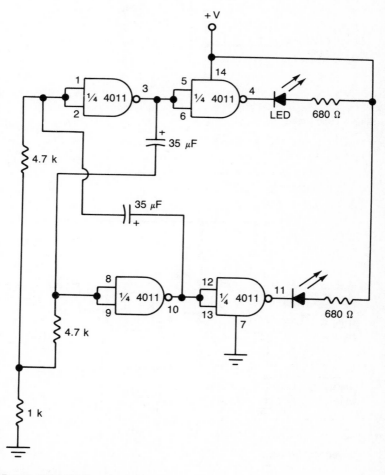

Double LED blinker

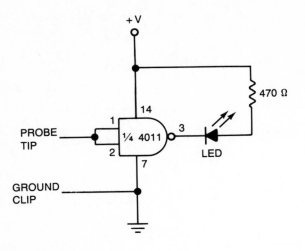

Simple logic probe

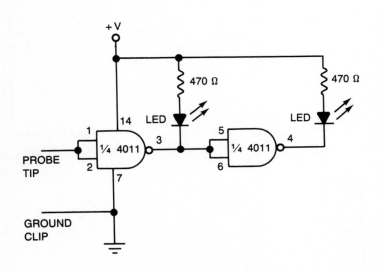

Improved logic probe

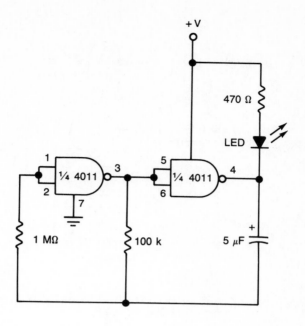

LED blinker

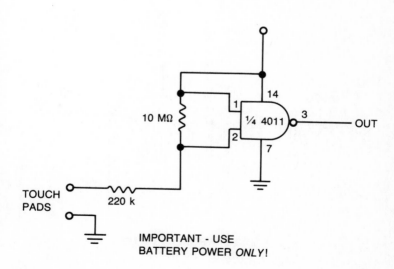

TOUCH
PADS

220 k

IMPORTANT - USE
BATTERY POWER *ONLY*!

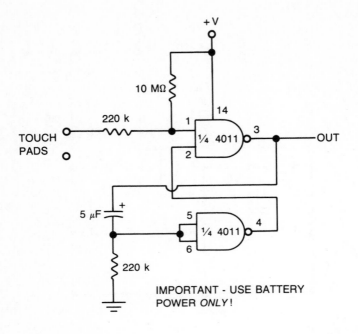

IMPORTANT - USE BATTERY
POWER *ONLY*!

Touch switch with time delay

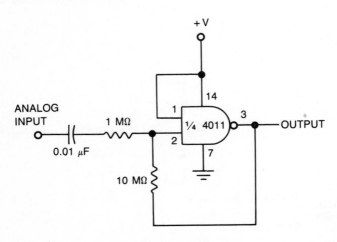

USING COMPONENT VALUES
SHOWN, GAIN IS 10

Linear amplifier

350

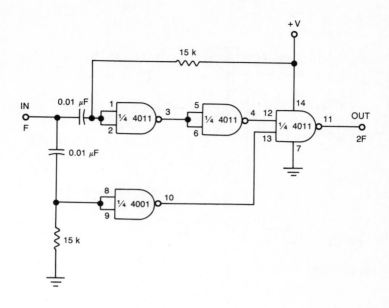

Frequency doubler

4012—CMOS DUAL FOUR INPUT NAND GATE

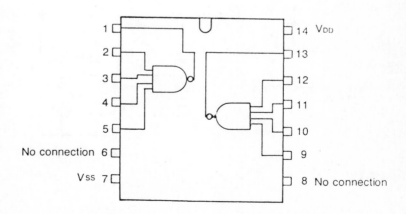

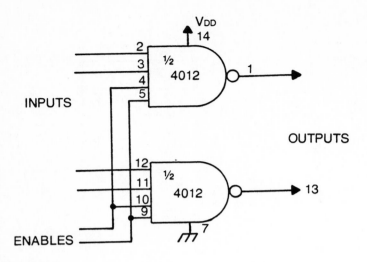

INPUTS

OUTPUTS

ENABLES

Enable input

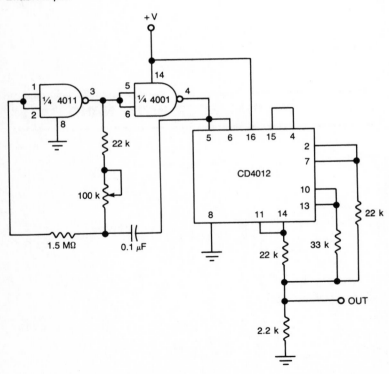

Four-stage stepped-wave generator

CD4013 DUAL D-TYPE FLIP FLOP

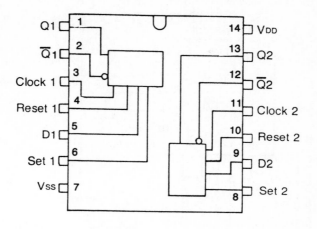

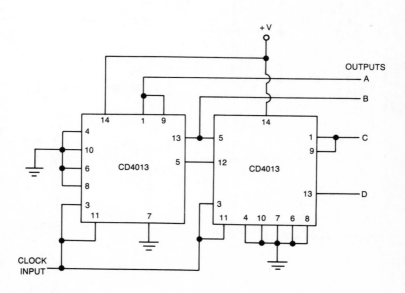

Eight-step binary counter

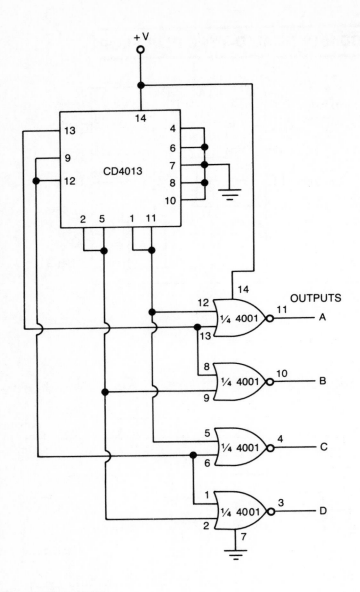

Four-step sequencer

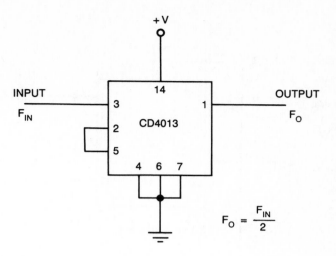

$$F_O = \frac{F_{IN}}{2}$$

Frequency halver

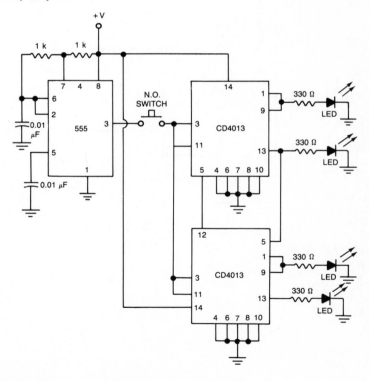

Random binary number generator

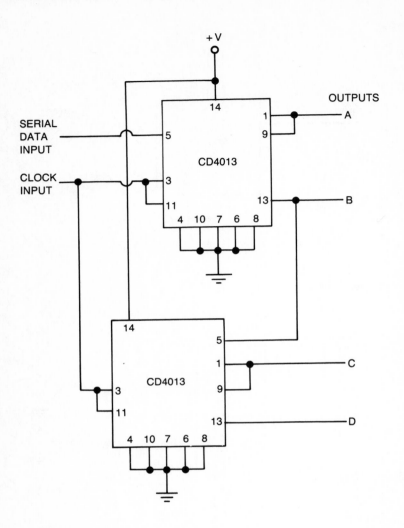

Shift register

356

4017—CMOS DECADE COUNTER/DECODER

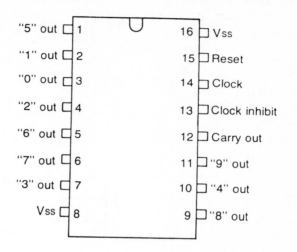

"5" out 1   16 Vss
"1" out 2   15 Reset
"0" out 3   14 Clock
"2" out 4   13 Clock inhibit
"6" out 5   12 Carry out
"7" out 6   11 "9" out
"3" out 7   10 "4" out
Vss 8   9 "8" out

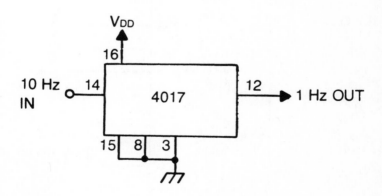

1-Hz timebase

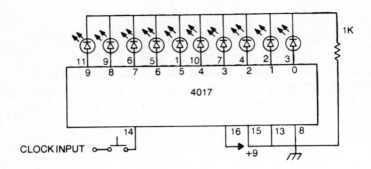

Random number generator

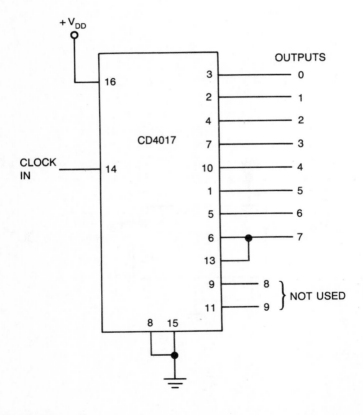

Count to seven and halt

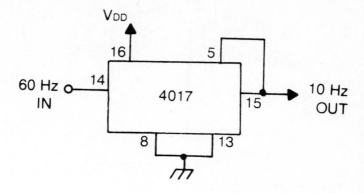

10-Hz timebase

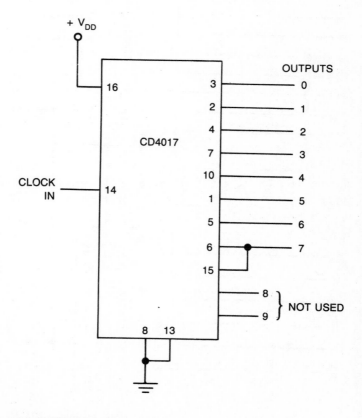

Count to seven and recycle

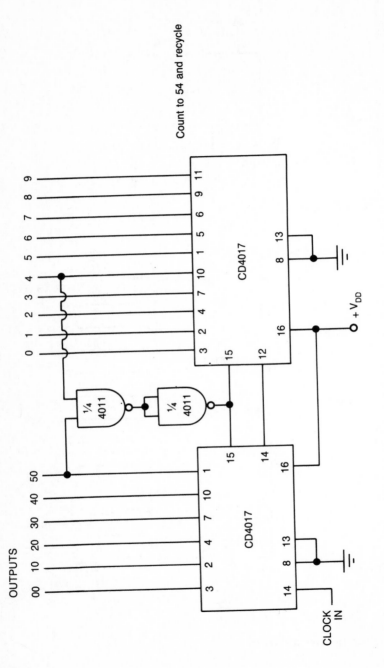

Count to 54 and recycle

360

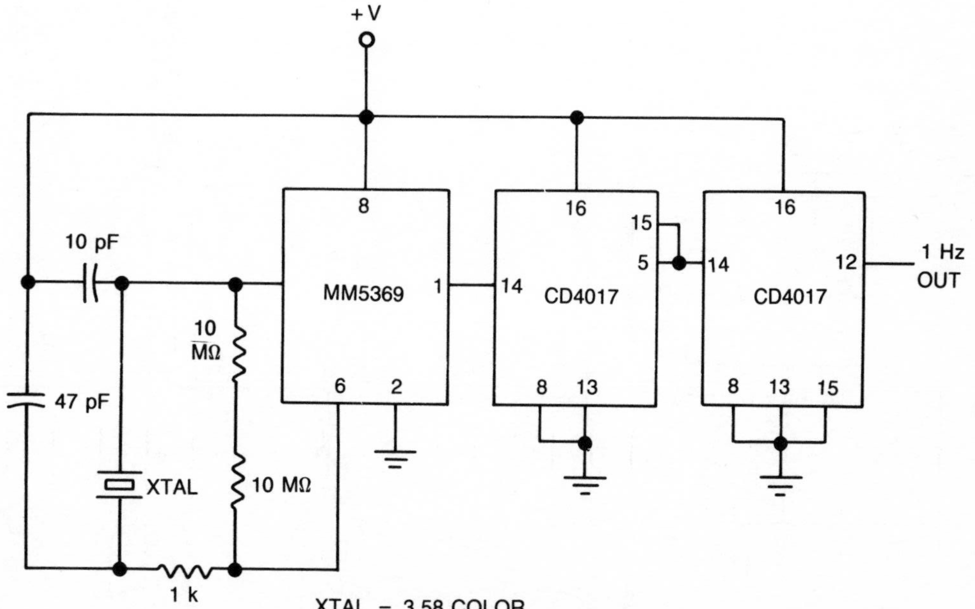

1 Hz OUT

12

CD4017

15

13

8

16

14

15

5

CD4017

16

13

8

14

1

MM5369

2

6

8

+V

10 MΩ

10 MΩ

1 k

10 pF

47 pF

XTAL

XTAL = 3.58 COLOR
BURST CRYSTAL

One hertz timebase

4023—CMOS TRIPLE
THREE INPUT NAND GATE

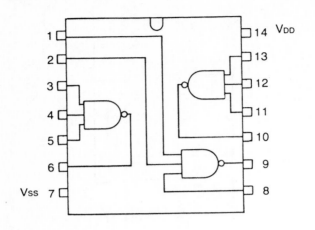

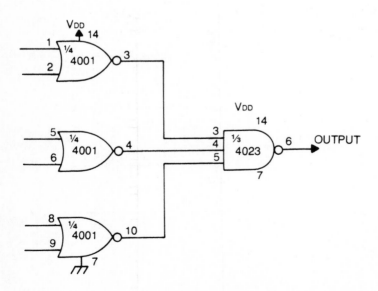

6-input OR gate

362

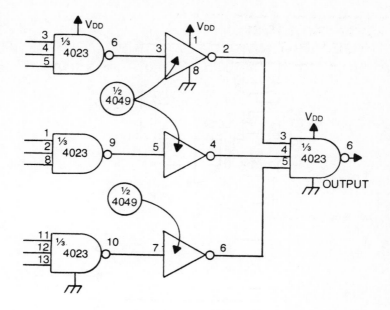

9-input NAND gate

4027—CMOS DUAL J-K FLIP-FLOP

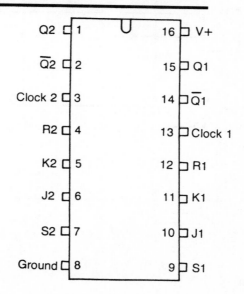

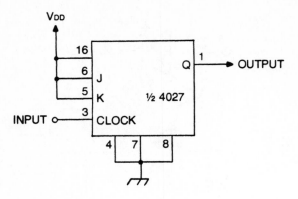

Divide-by-2 counter

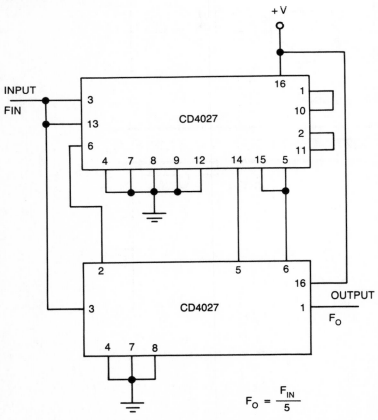

$$F_O = \frac{F_{IN}}{5}$$

Divide by five

364

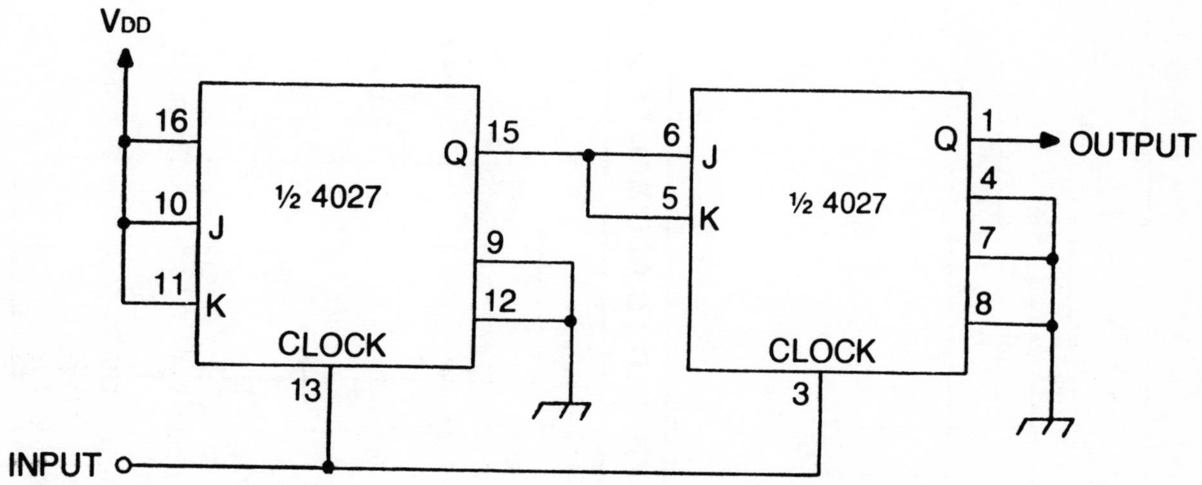

Divide-by-4 counter

365

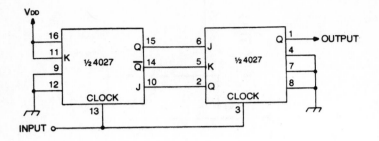

Divide-by-3 counter

4028—CMOS BCD TO DECIMAL DECODER

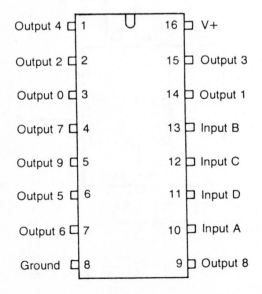

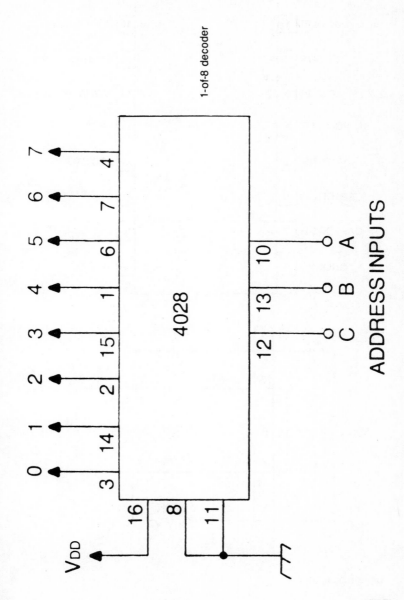

1-of-8 decoder

ADDRESS INPUTS

4046—CMOS PHASE
LOCK LOOP INTEGRATED CIRCUIT

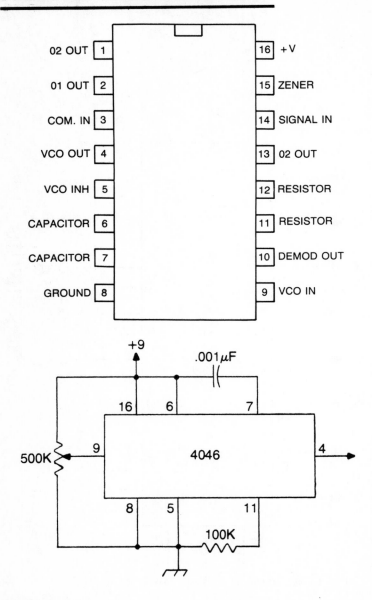

Tunable oscillator

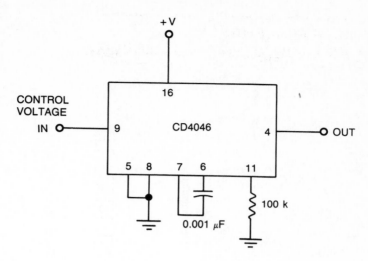

VCO

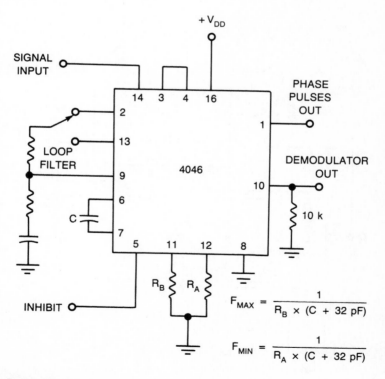

Phase locked loop

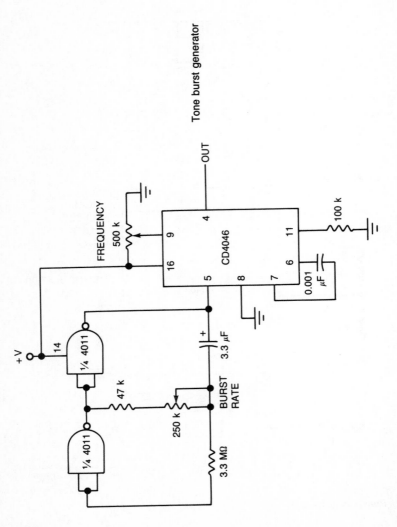

Tone burst generator

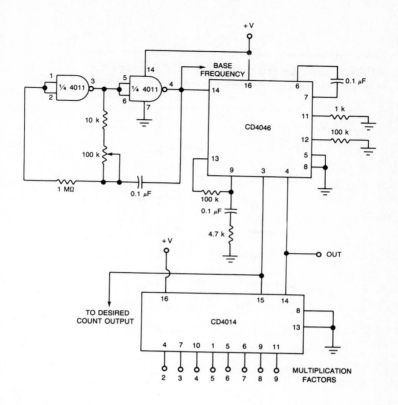

Frequency synthesizer

4049—CMOS HEX INVERTING BUFFER

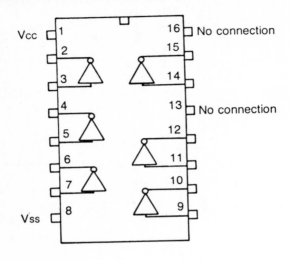

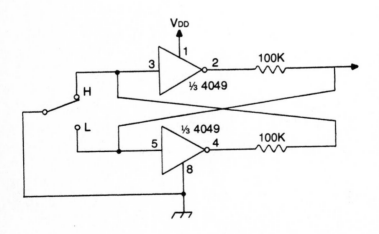

Bounceless switch

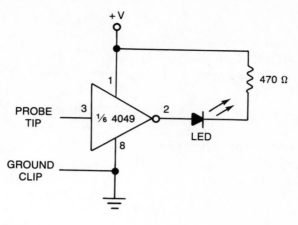

Logic probe

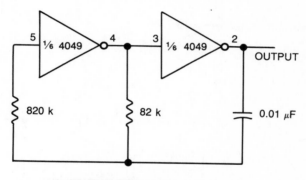

OUTPUT FREQUENCY ≅ 370 Hz

Clock generator

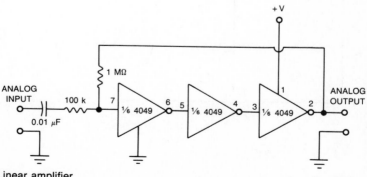

Linear amplifier

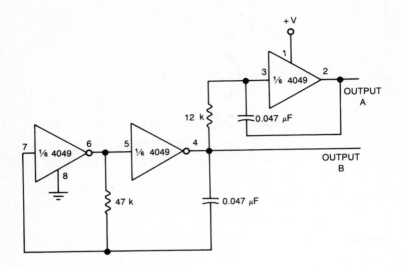

Two waveform function generator

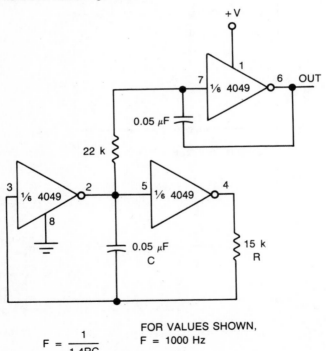

$$F = \frac{1}{1.4RC}$$

FOR VALUES SHOWN,
F = 1000 Hz

Triangle-wave generator

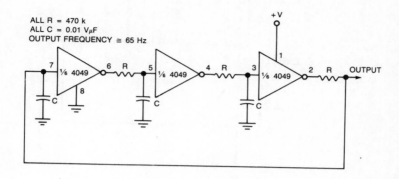

ALL R = 470 k
ALL C = 0.01 VμF
OUTPUT FREQUENCY $\cong$ 65 Hz

+V

Phase shift oscillator

4051—CMOS ANALOG MULTIPLEXER

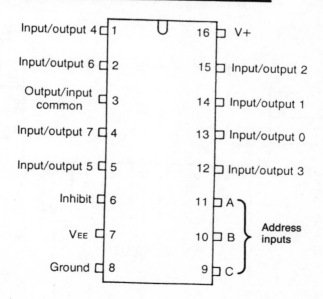

Input/output 4	1	16 V+
Input/output 6	2	15 Input/output 2
Output/input common	3	14 Input/output 1
Input/output 7	4	13 Input/output 0
Input/output 5	5	12 Input/output 3
Inhibit	6	11 A
V$_{EE}$	7	10 B
Ground	8	9 C

Address inputs

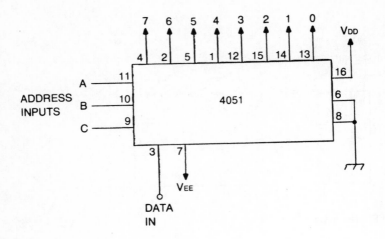

1-of-8 multiplexer

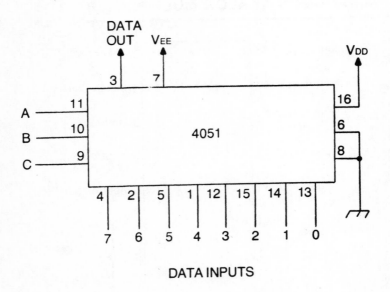

1-of-8 demultiplexer

4066—CMOS QUAD BILATERAL SWITCH

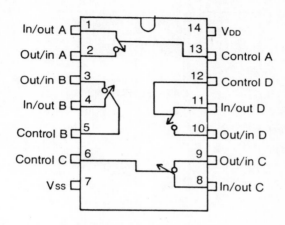

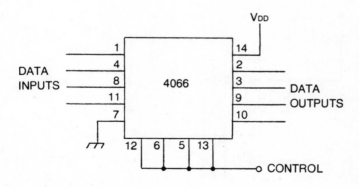

Data bus control

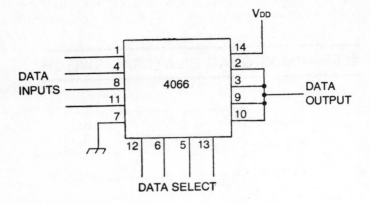

Data selector

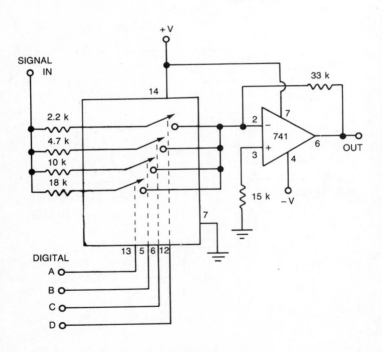

Inverting amplifier with programmable gain

378

4070—CMOS QUAD EXCLUSIVE OR GATE

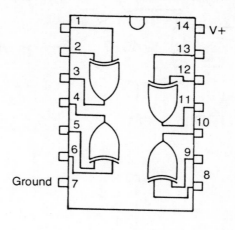

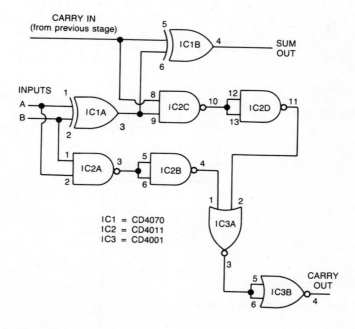

IC1 = CD4070
IC2 = CD4011
IC3 = CD4001

Binary adder

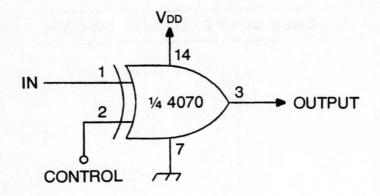

Controlled inverter

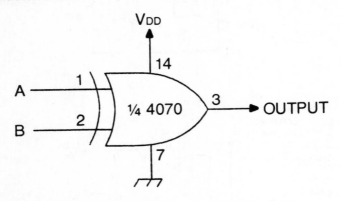

1-bit comparator

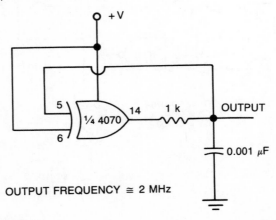

OUTPUT FREQUENCY ≅ 2 MHz

Clock generator

380

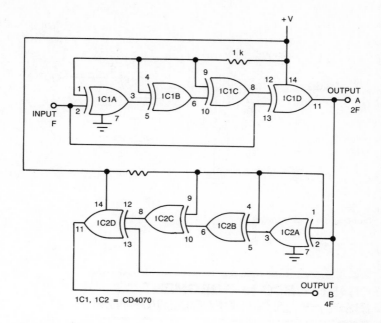

Frequency multiplier

1C1, 1C2 = CD4070

CD4081—QUAD 2-INPUT AND GATE

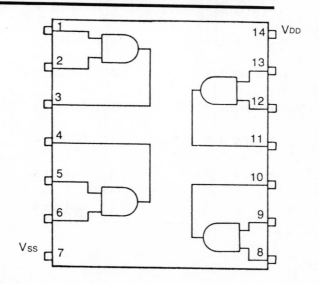

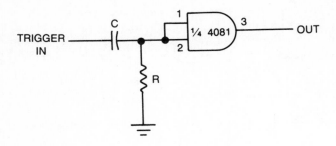

Monostable multivibrator

CD4511—BCD-to-7 SEGMENT
DISPLAY LATCH/DECODER/DRIVER

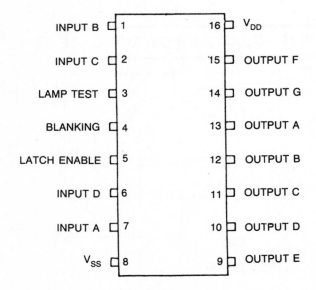

INPUT B 1	16 V_{DD}

INPUT B — 1   16 — V_{DD}

INPUT C — 2   15 — OUTPUT F

LAMP TEST — 3   14 — OUTPUT G

BLANKING — 4   13 — OUTPUT A

LATCH ENABLE — 5   12 — OUTPUT B

INPUT D — 6   11 — OUTPUT C

INPUT A — 7   10 — OUTPUT D

V_{SS} — 8   9 — OUTPUT E

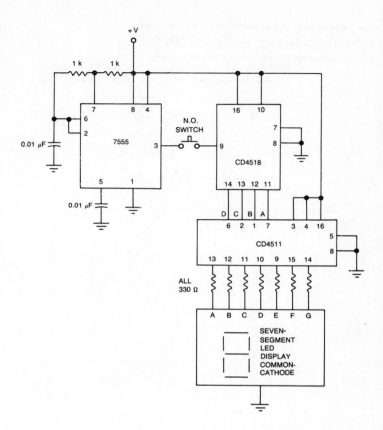

Random number generator

CD4518—DUAL BCD COUNTER

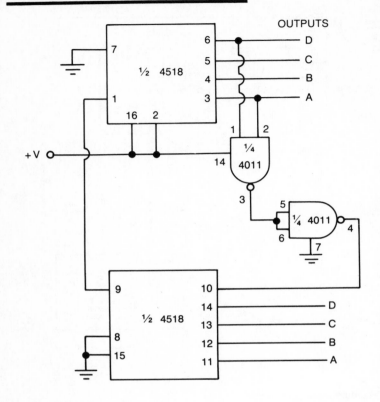

Cascaded BCD counters

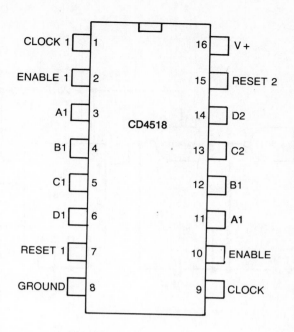

CD4528—DUAL ONE SHOT

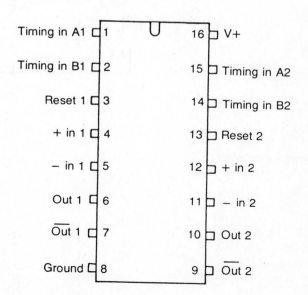

385

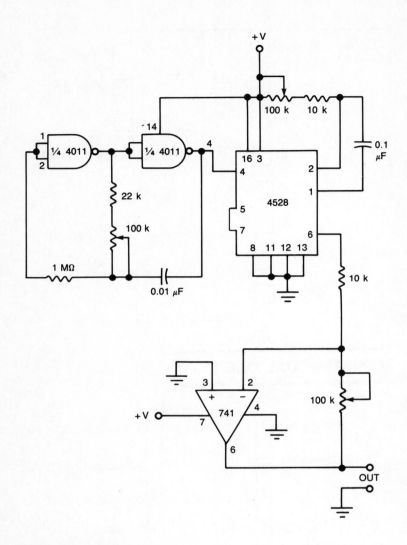

Stepped-wave generator

7555—CMOS TIMER

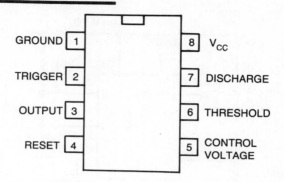

THE 7555 CMOS TIMER

The 7555 is a CMOS version of the 555 integrated timer (see Part 2 of this book). It can be used in all of the same applications, but it is designed specifically to be used with CMOS digital circuits. The standard 555 can also be used directly with CMOS devices, but the 7555 offers several advantages, including significantly lower power consumption, and a wider range of acceptable power supply voltages ($+2$ to $+18$ volts). The CMOS circuitry within the 7555 chip also permits longer timing periods without the instability problems that can occur with the standard 555 timer.

Like the 555, the 7555 may be operated in either the monostable or the astable mode. The basic timing formula is the same as for the 555;

$$T = 1.1RC$$

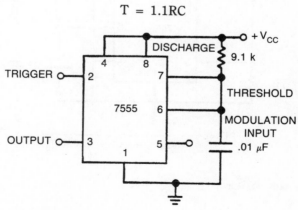

Pulse width modulator

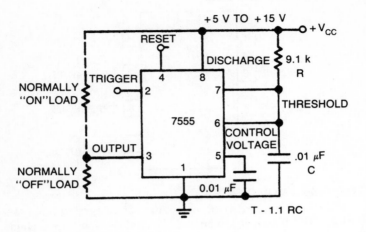

Monostable multivibrator

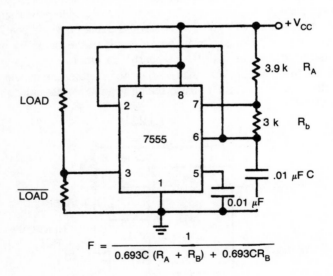

$$F = \frac{1}{0.693C (R_A + R_B) + 0.693CR_B}$$

Astable multivibrator

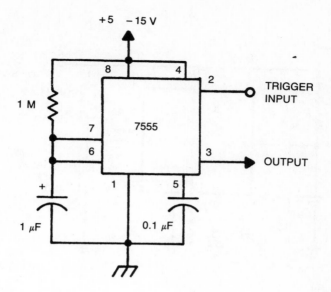

One-shot timer

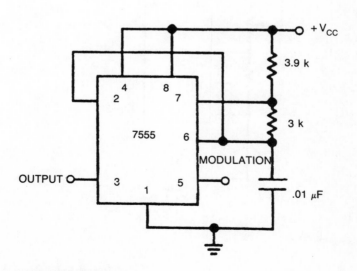

Pulse position modulator

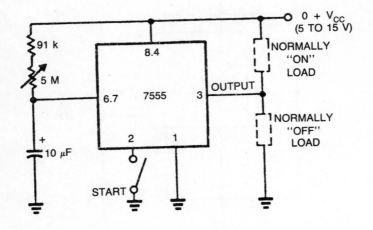

On or off control

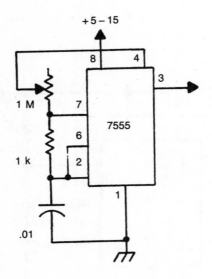

Pulse generator

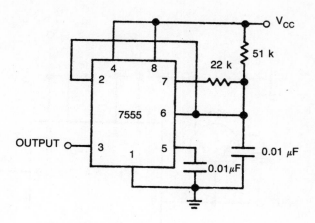

50% duty cycle oscillator

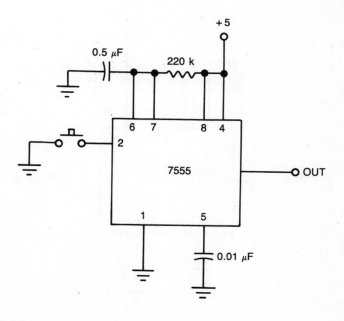

Switch debouncer

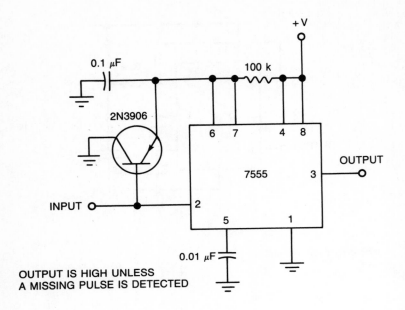

OUTPUT IS HIGH UNLESS
A MISSING PULSE IS DETECTED

Missing pulse detector

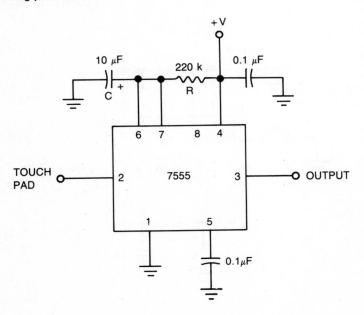

Timed touch switch

392

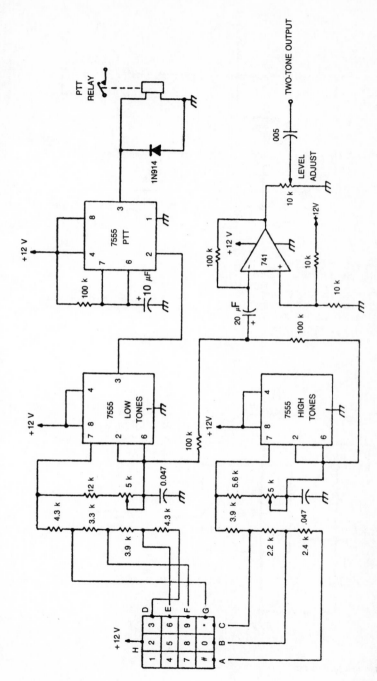

Telephone dialing tone encoder

393

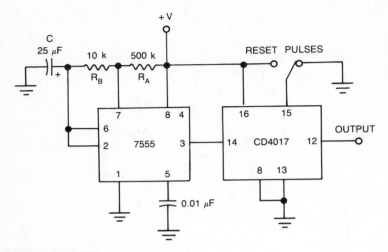

VERY LOW RATE
PULSE GENERATOR

ADD MORE CD4017 FOR SLOWER
PULSE RATES. EACH CD4017 DIVIDES
THE INPUT FREQUENCY BY TEN.

$$F = \frac{1}{6.93C \, (R_A + R_B) + 6.93CR_B}$$

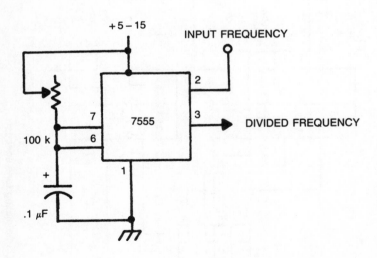

Frequency divider

394

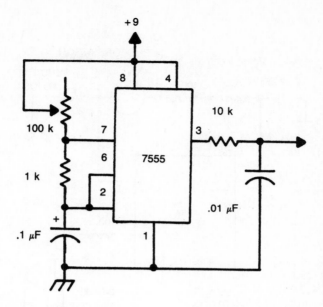

Triangle-wave generator

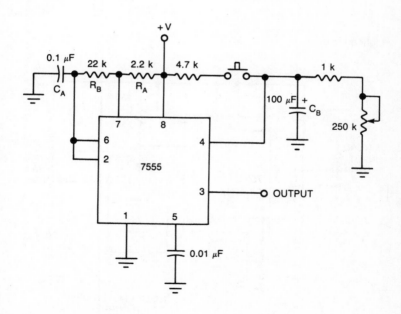

Tone burst generator

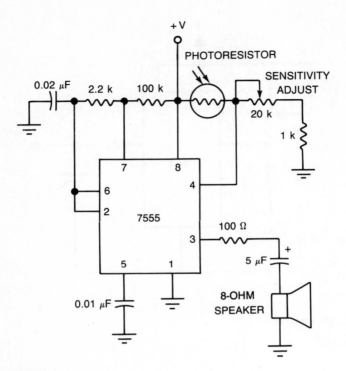

Light on alarm

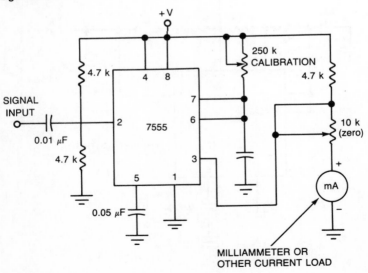

Frequency meter

Event failure alarm

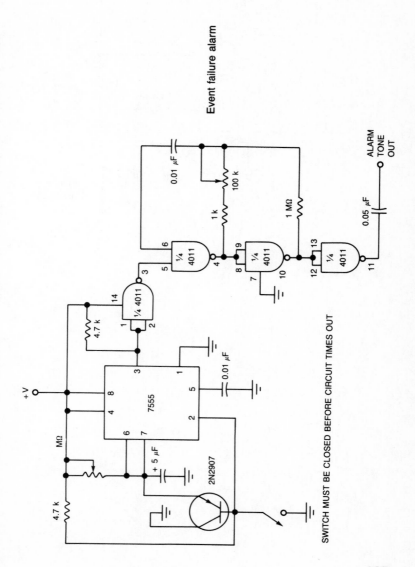

SWITCH MUST BE CLOSED BEFORE CIRCUIT TIMES OUT

397

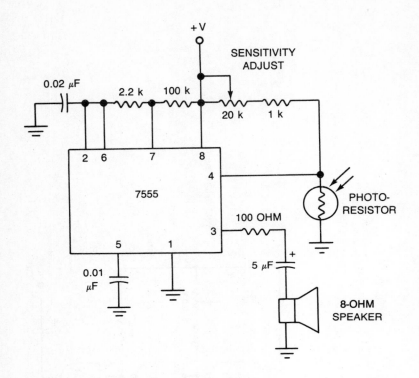

Light off alarm

Part 5
TTL
Integrated Circuits

The so-called "digital revolution" in electronics really got its start with the development of the TTL logic family. It was the first practical and popular class of digital ICs.

TTL devices range from simple gates (SSI — Small Scale Integration), up to moderately complex (MSI — Medium Scale Integration) devices, such as counters and multiplexers. TTL is not really well suited to Large Scale (LSI) or Very Large Scale Integration (VLSI) devices.

The standard numbering scheme for TTL ICs is in the form of 74xx. Military-grade devices are numbered 54xx. For all but the most critical applications (generally in terms of ambient temperature range), 54xx and 74xx devices are 100% interchangeable.

In the 74xx numbering scheme, "xx" is a two or three digit number uniquely identifying the specific device. The same numbering scheme is used for a variety of TTL subfamilies, and even a line of CMOS units. Two devices with the same ID number (xx) have identical functions and pinouts, regardless of the subfamily. Devices in different subfamilies, however, may not be interchangeable because of differences in power supply and signal level requirements. The subfamily is indicated by one or two letters in the middle of the type number;

74xx	Standard TTL
74Lxx	Low-Power TTL

74Hxx	High-Speed TTL
74LSxx	Low-Power Schottky TTL
74Cxx	CMOS

Any of the subfamilies may be used in any of the applications in this section, as long as the appropriate supply and signal conventions are followed.

7400—TTL QUAD NAND GATE

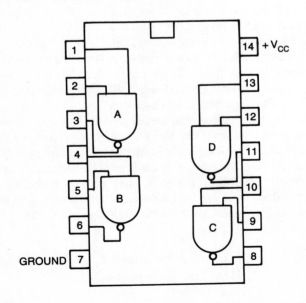

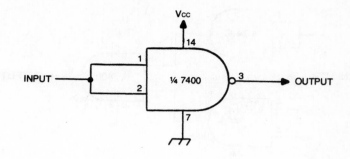

Inverter

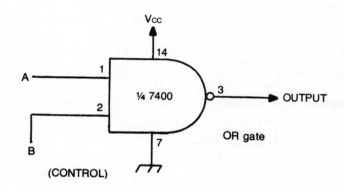

Control gate

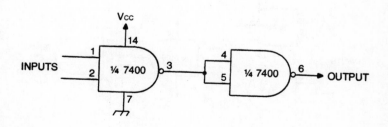

AND gate

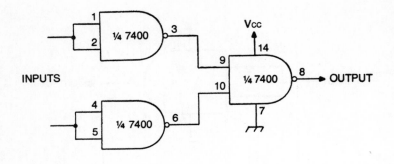

OR gate

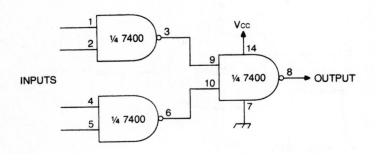

AND-OR gate

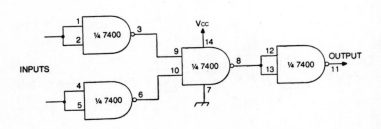

NOR gate

402

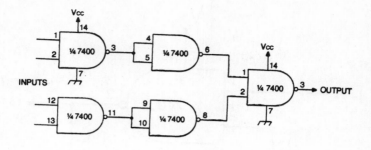

4-Input NAND gate

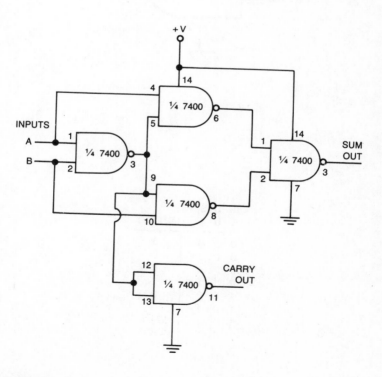

Half-adder

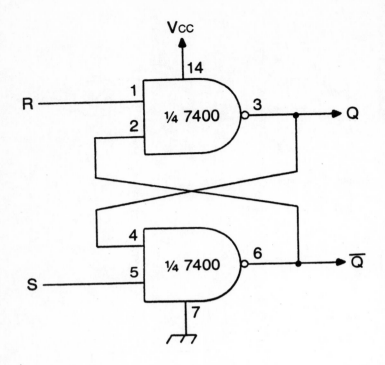

RS latch

INPUTS			OUTPUT
A	B	C	
0	0	0	0
0	0	1	0
0	1	0	0
0	1	1	1
1	0	0	0
1	0	1	1
1	1	0	1
1	1	1	1

Truth table for majority logic circuit

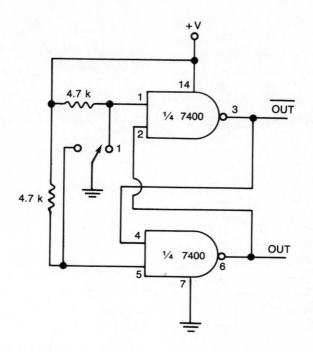

Bounceless switch

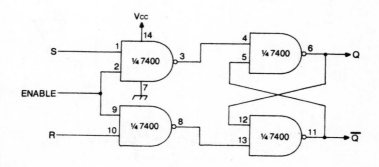

Gated RS latch

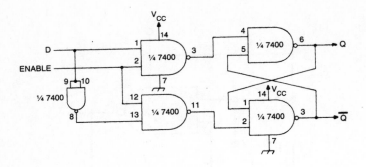

D flip-flop

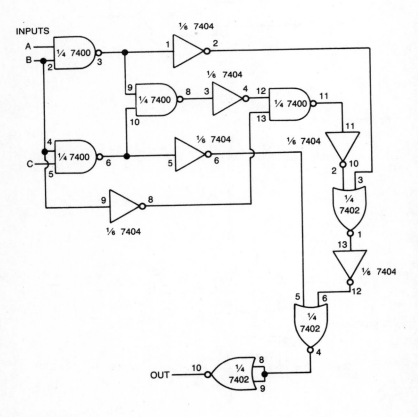

Majority logic

406

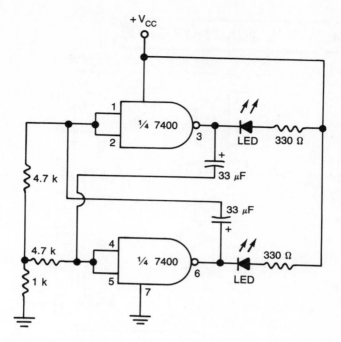

Dual LED blinker

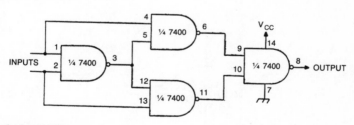

Exclusive-NOR gate

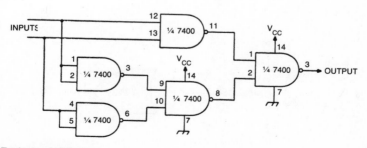

Exclusive-NOR gate

7402—TTL QUAD NOR GATE

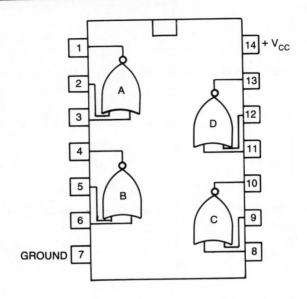

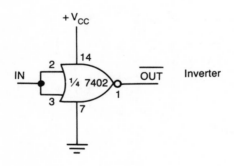

Inverter

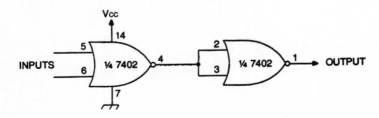

OR gate

408

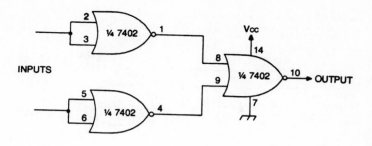

AND gate

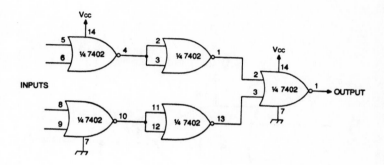

4-Input NOR gate

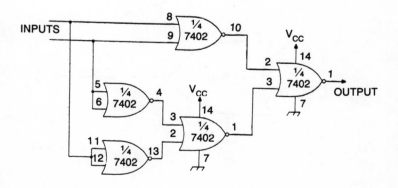

Exclusive-OR gate

409

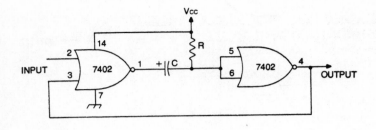

One-shot

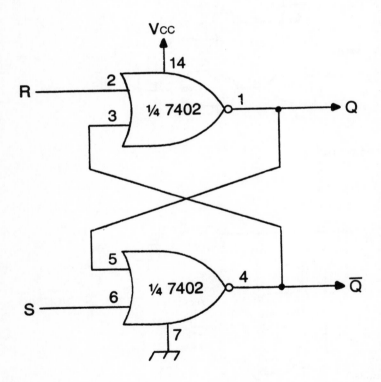

RS latch

410

7404—HEX INVERTER

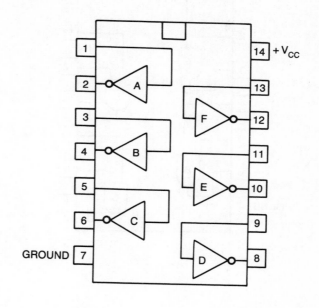

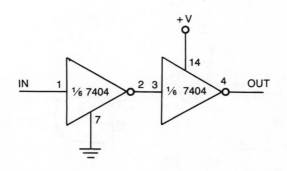

Buffer

7408—QUAD AND GATE

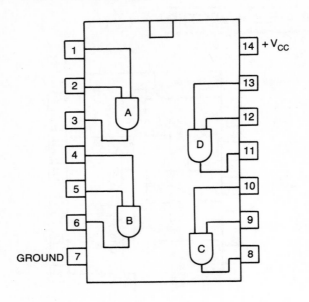

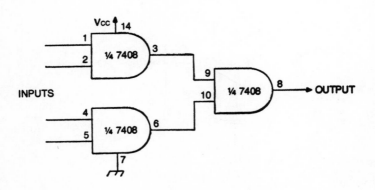

4-input AND gate

412

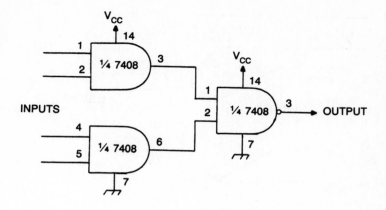

4-input NAND gate

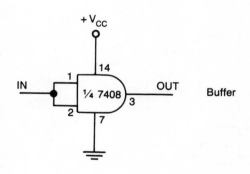

Buffer

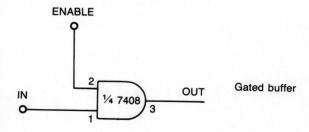

Gated buffer

413

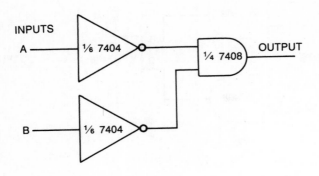

NOR gate

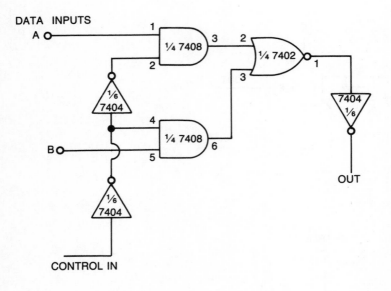

2 Input data selector

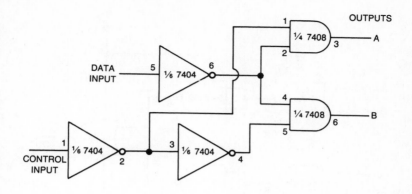

1-of-2 demultiplexer

7432—QUAD OR GATE

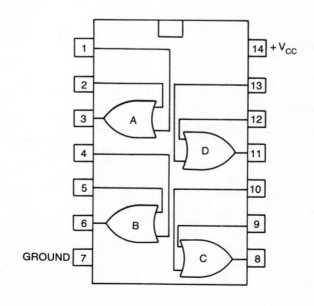

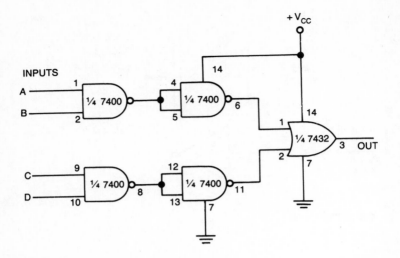

AND/OR gate

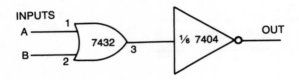

NOR gate

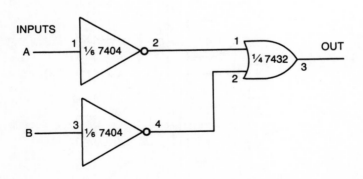

NAND gate

416

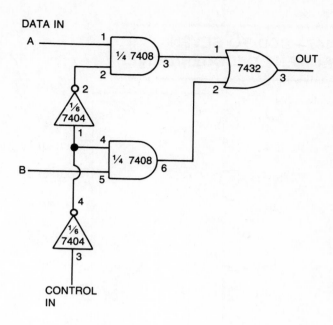

DATA IN

2 Input data selector

7447—BCD TO SEVEN-SEGMENT DECODER/DRIVER

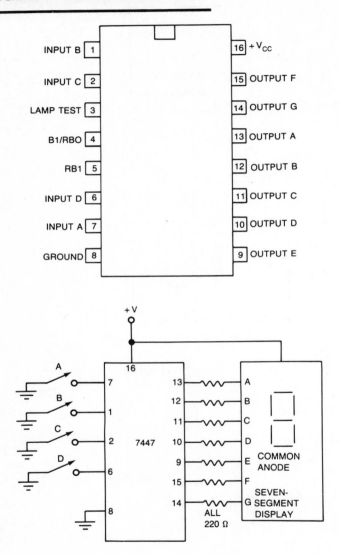

Manual BCD/7-segment demonstrator

418

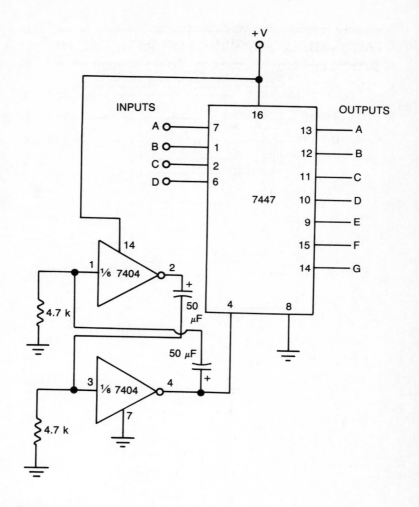

Flashing display

7473—DUAL J-K FLIP-FLOP WITH CLEAR

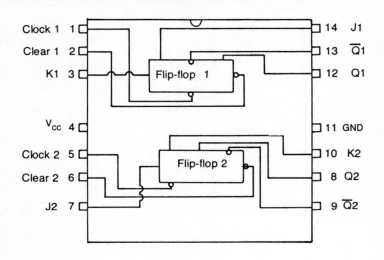

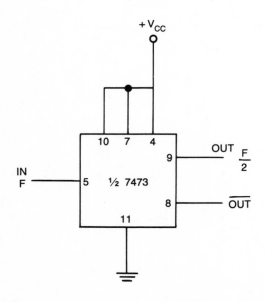

Divide by two

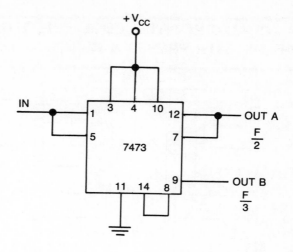

Divide by three

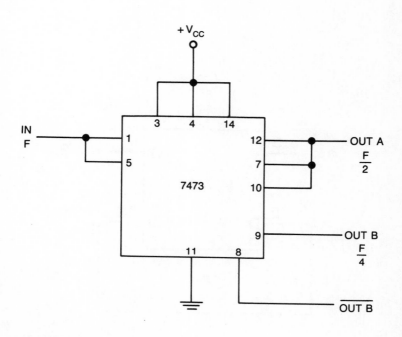

Divide by four

421

7474—DUAL D POSITIVE-EDGE TRIGGERED FLIP-FLOP WITH PRESET & CLEAR

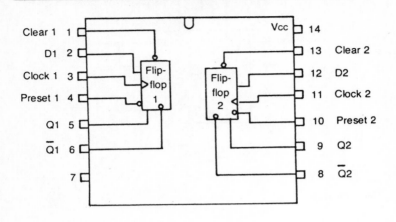

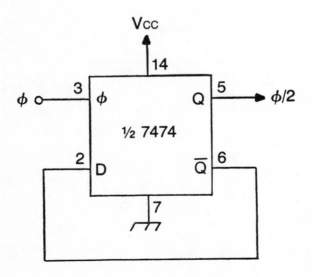

Divide-by-two counter

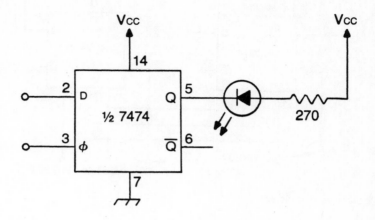

Phase detector

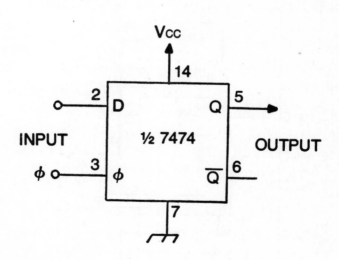

Wave shaper

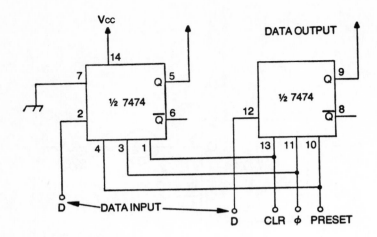

2-bit storage register

7475—QUAD LATCH

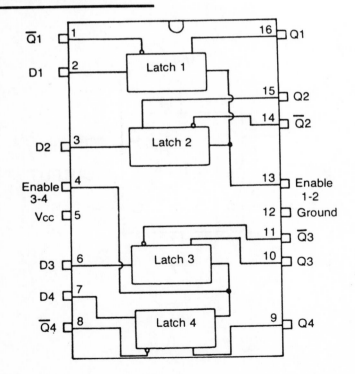

424

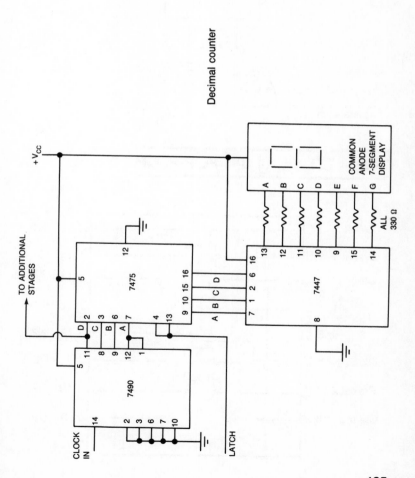

Decimal counter

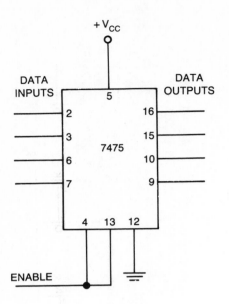

DATA INPUTS — 2, 3, 6, 7
+ V_CC — 5
7475
DATA OUTPUTS — 16, 15, 10, 9
ENABLE — 4, 13
12 — GND

4-bit data latch

7476—TTL DUAL J-K FLIP-FLOP

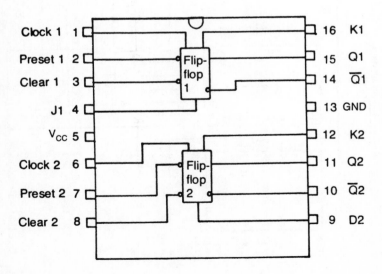

Clock 1 1	16 K1
Preset 1 2	15 Q1
Clear 1 3	14 $\overline{Q1}$
J1 4	13 GND
V_CC 5	12 K2
Clock 2 6	11 Q2
Preset 2 7	10 $\overline{Q2}$
Clear 2 8	9 D2

Flip-flop 1

Flip-flop 2

426

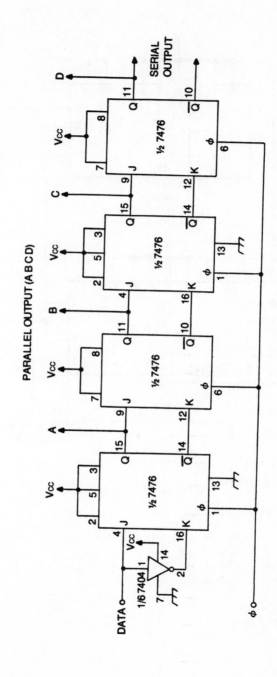

4-bit serial shift register

427

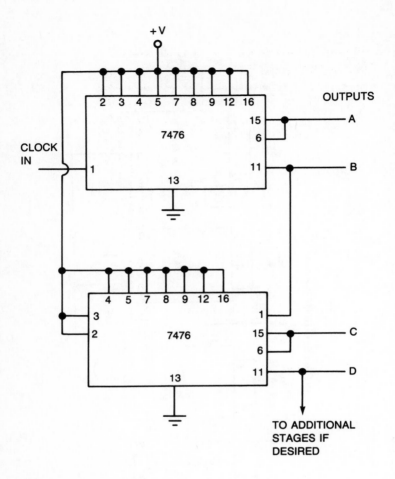

4-bit binary counter

428

7490—TTL BCD DECADE COUNTER

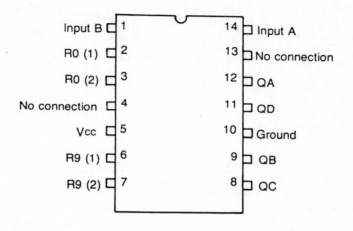

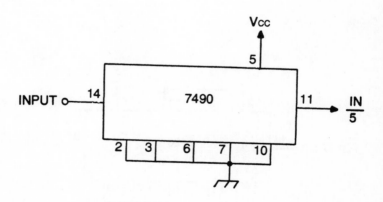

Divide-by-5 counter

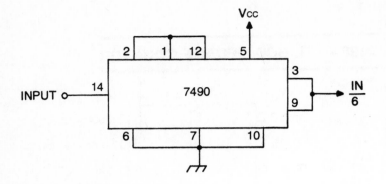

Divide-by-6 counter

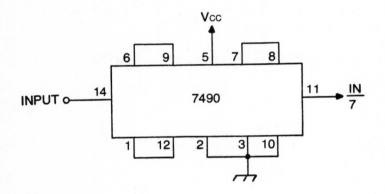

Divide-by-7 counter

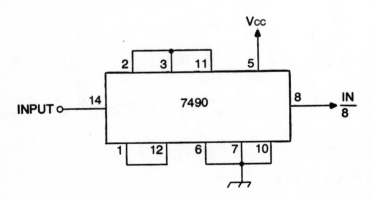

Divide-by-8 counter

430

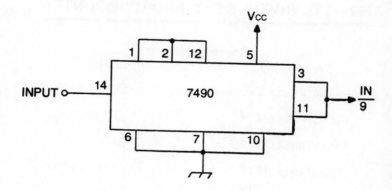

Divide-by-9 counter

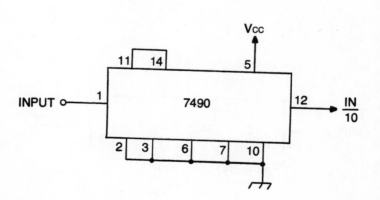

Divide-by-10 counter

7492—TTL DIVIDE BY 12 BINARY COUNTER

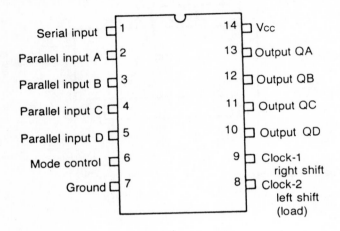

Serial input — 1 — 14 — Vcc
Parallel input A — 2 — 13 — Output QA
Parallel input B — 3 — 12 — Output QB
Parallel input C — 4 — 11 — Output QC
Parallel input D — 5 — 10 — Output QD
Mode control — 6 — 9 — Clock-1 right shift
Ground — 7 — 8 — Clock-2 left shift (load)

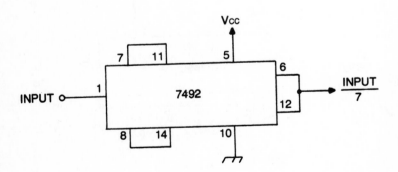

Divide-by-7 counter

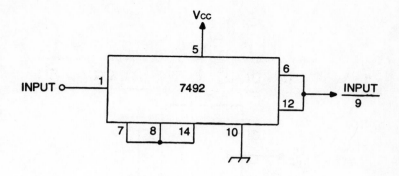

Divide-by-9 counter

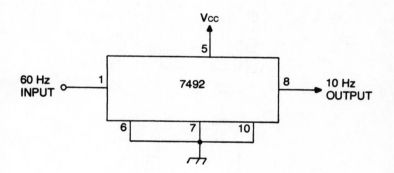

10-Hz pulse source

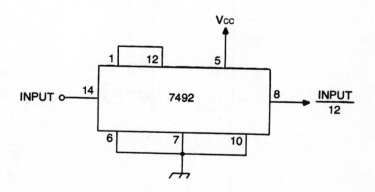

Divide-by-12 counter

Divide-by-120 counter

74123—DUAL
RETRIGGERABLE ONE SHOT WITH CLEAR

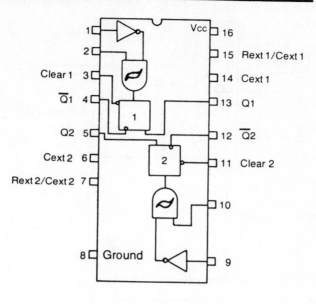

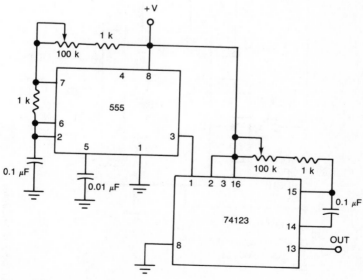

Sound effect generator

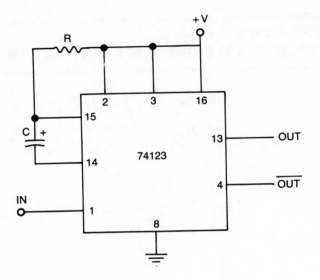

Missing pulse detector

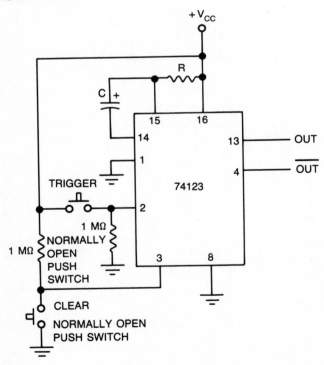

One shot

436

74132—QUAD NAND SCHMITT TRIGGER

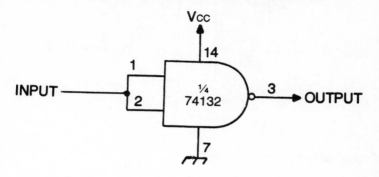

Noise eliminator

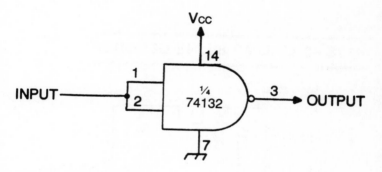

Threshold detector

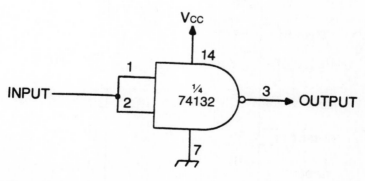

Wave shaper

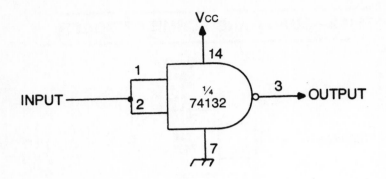

Pulse restorer

74138—3 LINE TO 8 LINE DECODER

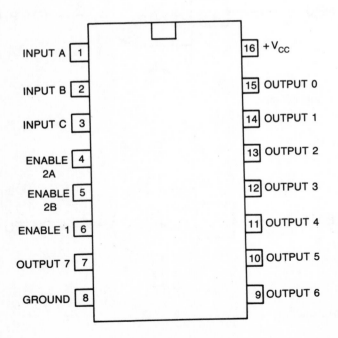

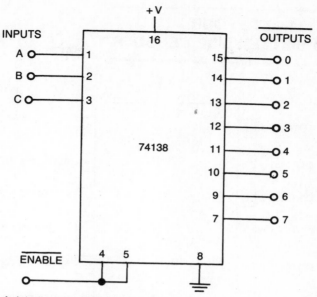

Three-of-eight line decoder/multiplexer

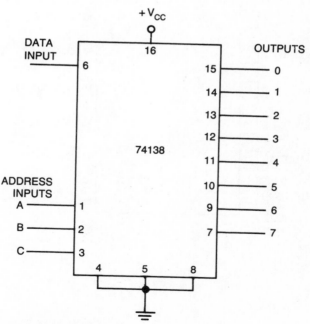

One-of-eight multiplexer

439

74154—TTL FOUR TO SIXTEEN LINE DECODER

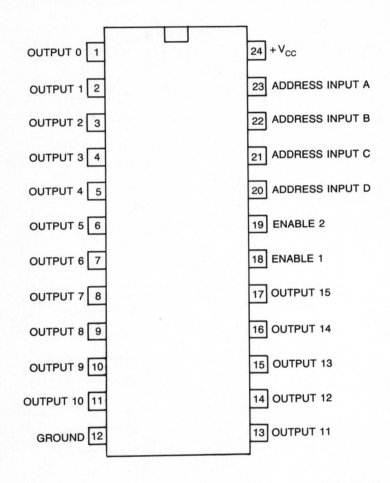

OUTPUT 0 — 1   24 — +V_{CC}

OUTPUT 0 [1]	[24] +V$_{CC}$
OUTPUT 1 [2]	[23] ADDRESS INPUT A
OUTPUT 2 [3]	[22] ADDRESS INPUT B
OUTPUT 3 [4]	[21] ADDRESS INPUT C
OUTPUT 4 [5]	[20] ADDRESS INPUT D
OUTPUT 5 [6]	[19] ENABLE 2
OUTPUT 6 [7]	[18] ENABLE 1
OUTPUT 7 [8]	[17] OUTPUT 15
OUTPUT 8 [9]	[16] OUTPUT 14
OUTPUT 9 [10]	[15] OUTPUT 13
OUTPUT 10 [11]	[14] OUTPUT 12
GROUND [12]	[13] OUTPUT 11

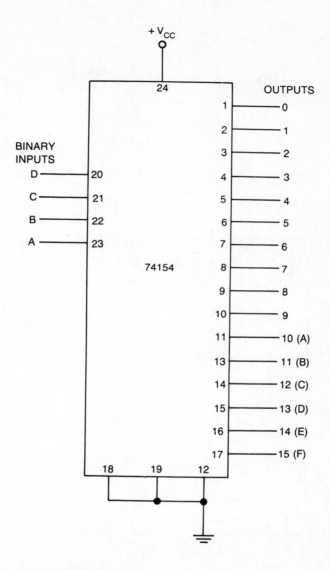

Binary to hexadecimal converter

441

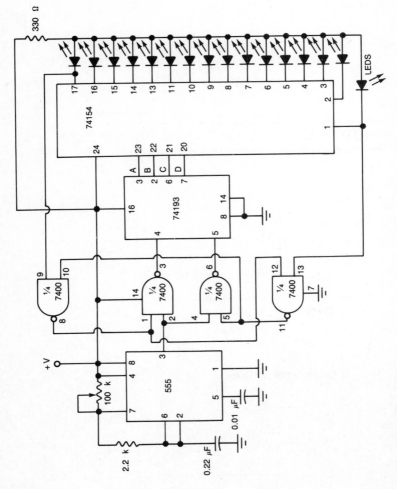

330 Ω

LEDS

74154

74193

555

1/4 7400

100 k

2.2 k

0.22 μF

0.01 μF

+V

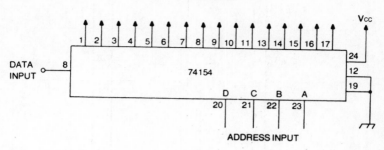

1-of-16 demultiplexer

74192—TTL BCD UP/DOWN COUNTER

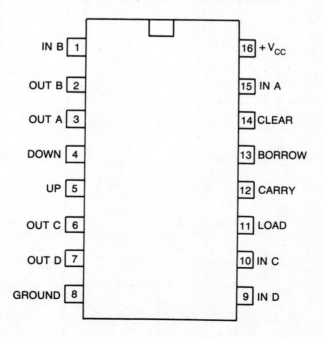

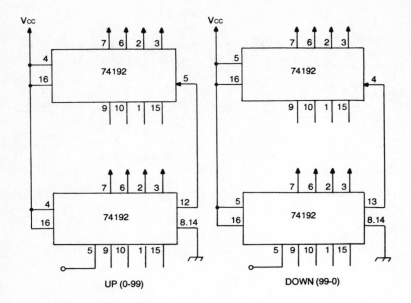

Cascaded counters

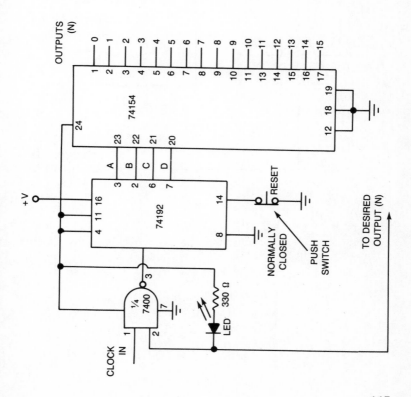

Count up to N and stop

445

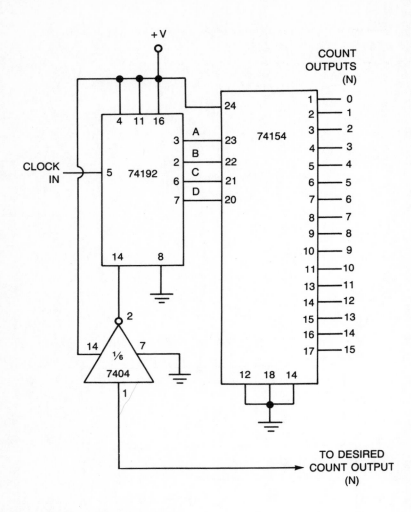

Count up to N and recycle

74193—TTL FOUR BIT UP/DOWN COUNTER

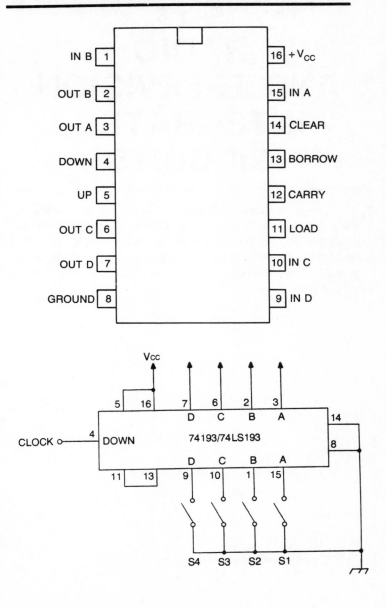

Count down from N and recycle

Part 6
RADIO AND TELEVISION INTEGRATED CIRCUITS

Not all chips wind up in audio amplifiers or logic circuits. Numerous ICs have been developed for use in radio and television applications. Many of these are unknown to the casual electronics experimenter. However, they are now becoming available on the surplus electronics market. This section will look at how some of these ICs can be used.

172—INTEGRATED AM IF STRIP

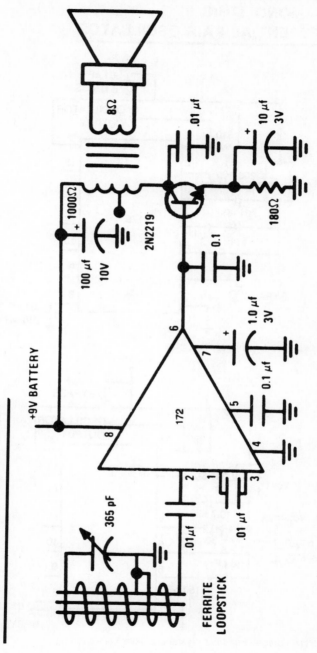

T.R.F. broadcast receiver

449

175—MONOLITHIC
DIFFERENTIAL PAIR OSCILLATOR

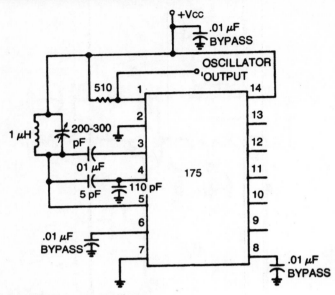

10-MHz L-C sine-wave oscillator

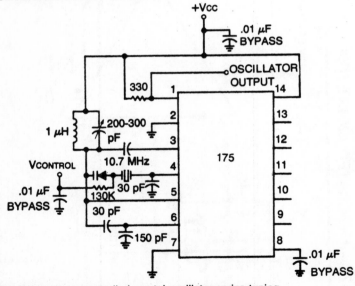

10.7-MHz voltage controlled crystal oscillator series tuning

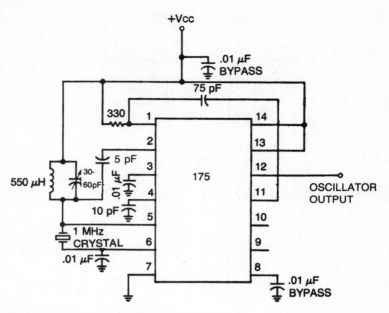

1-MHz crystal oscillator with TTL output

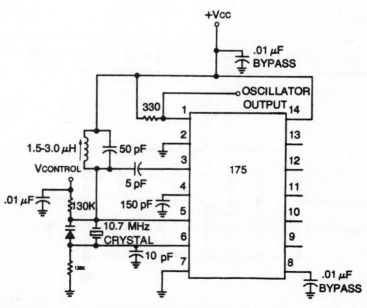

10.7-MHz voltage-controlled crystal oscillator parallel tuning

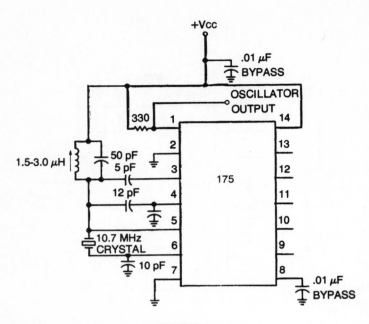

10.7-MHz parallel resonant crystal oscillator

703—LOW POWER DRAIN RF/IF AMPLIFIER

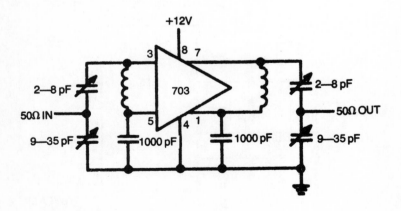

100-MHz narrow band amplifier

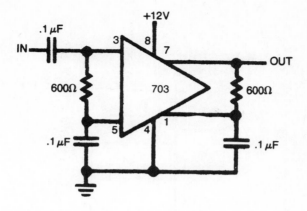

RC coupled video amplifier

1808—MONOLITHIC
TELEVISION SOUND SYSTEM

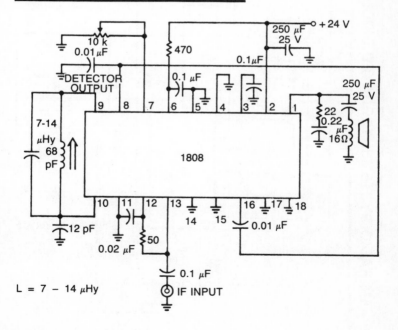

L = 7 – 14 µHy

Television sound system

720—AM RADIO SYSTEM INTEGRATED CIRCUIT

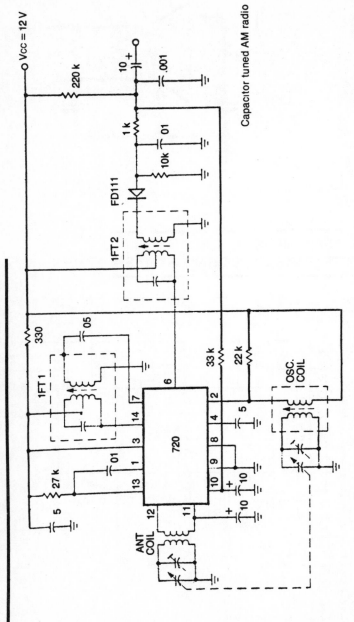

732—FM STEREO MULTIPLEX DECODER

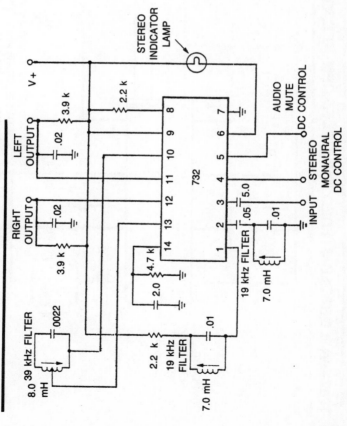

FM stereo multiplex decoder circuit

455

1307—INTEGRATED STEREO MULTIPLEX DEMODULATOR

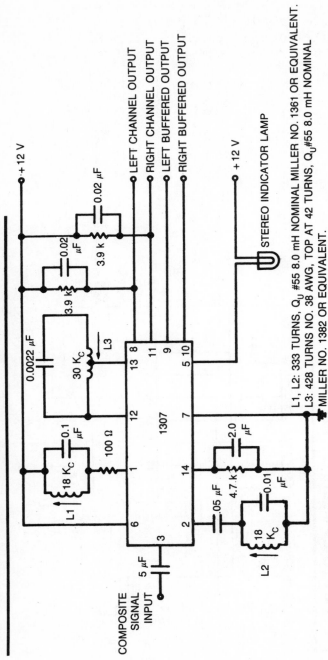

Stereo multiplex demodulator with beacon lamp

1310—PHASE-LOCKED LOOP FM STEREO DEMODULATOR

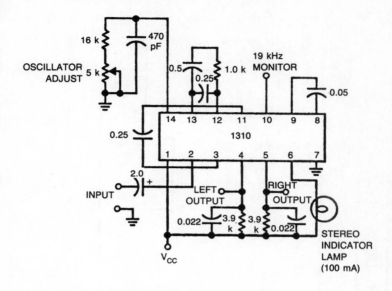

FM stereo demodulator with indicating lamp

1349—INTEGRATED IF AMPLIFIER WITH AGC

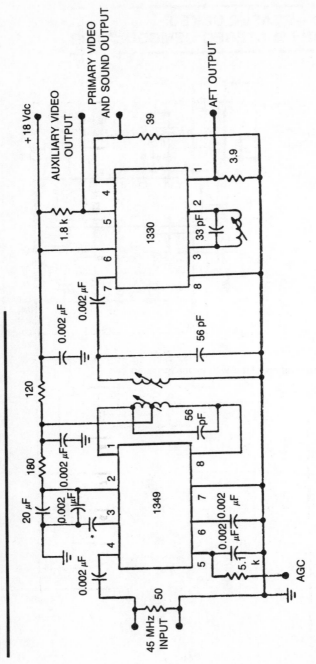

Video IF amplifier

458

1350—INTEGRATED AMPLIFIER

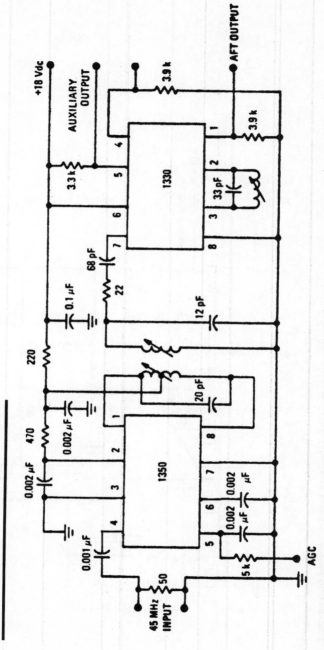

Video IF amplifier

459

1351—INTEGRATED TV SOUND CIRCUIT

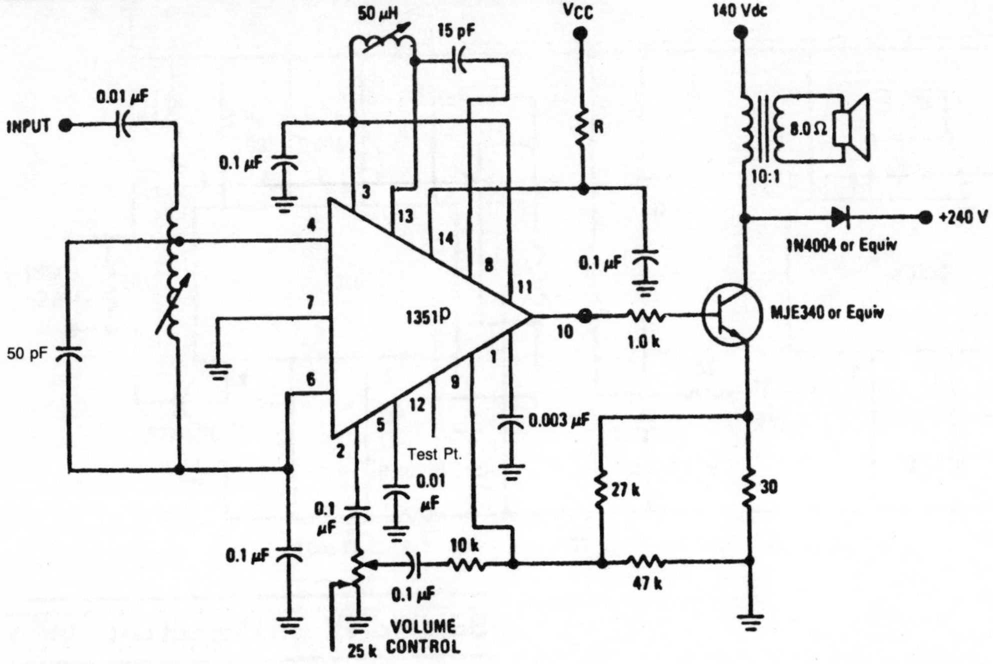

4.5-MHz typical application

460

1352—TV VIDEO AMPLIFIER WITH AGC

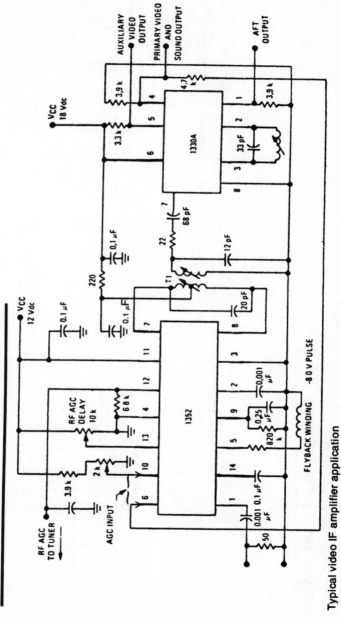

Typical video IF amplifier application

461

1355—BALANCED FOUR STAGE HIGH GAIN FM/IF AMPLIFIER

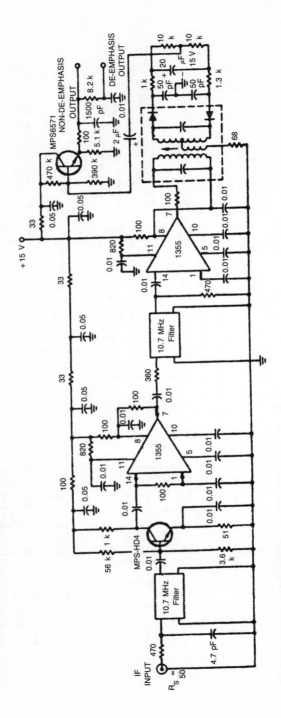

Dual 1355 FM IF application

1357—INTEGRATED IF AMPLIFIER AND QUADRATURE DETECTOR

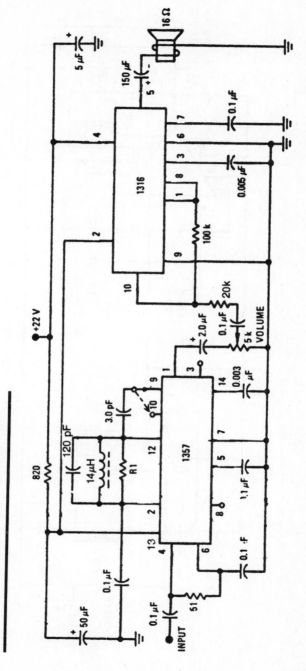

TV typical application circuit

463

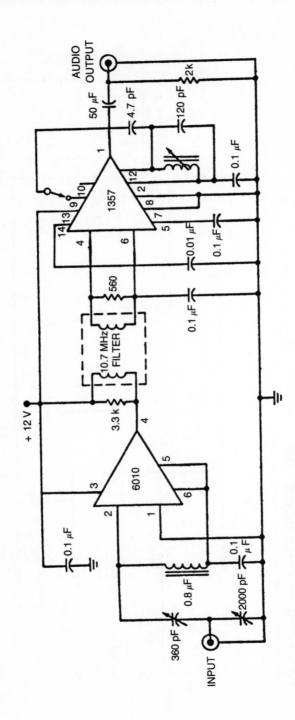

FM radio typical application circuit

1358—INTEGRATED TELEVISION SOUND IF AMPLIFIER

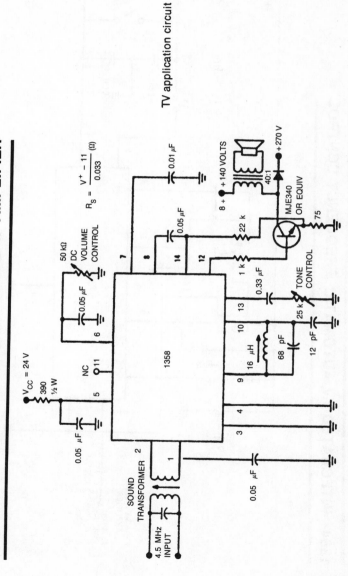

$$R_S = \frac{V^+ - 11}{0.033} \, (\Omega)$$

TV application circuit

465

1364—INTEGRATED TV AUTOMATIC FREQUENCY CONTROL

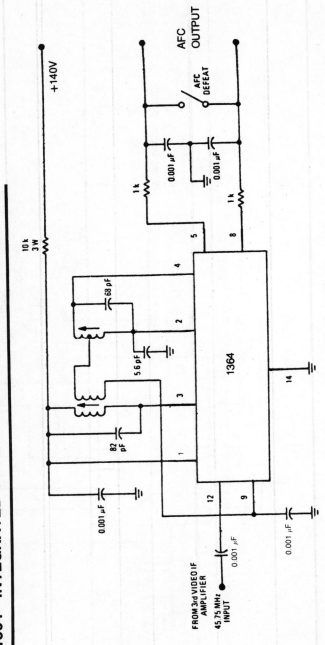

Typical AFC application

1301—INTEGRATED PHASE-LOCK LOOP

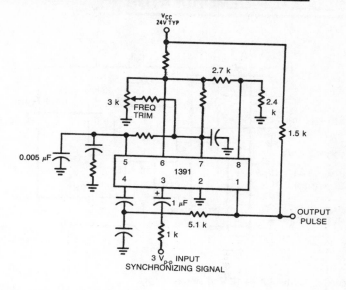

General purpose phase-lock loop

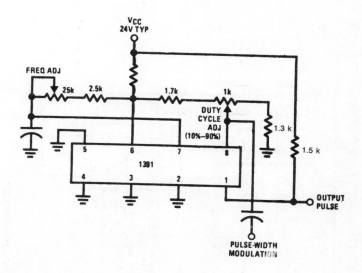

Variable duty cycle oscillator

1496—BALANCED
MODULATOR/DEMODULATOR

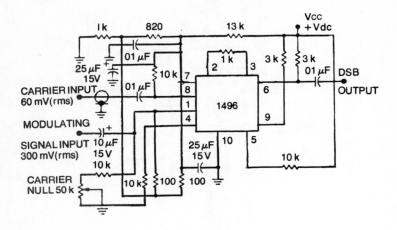

Balanced modulator (+12 Vdc single supply)

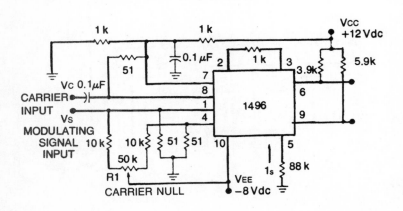

Balanced modulator/demodulator

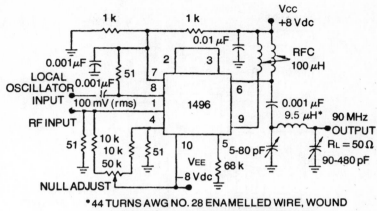

Doubly balanced mixer (broadband inputs, 9.0-MHz tuned output)

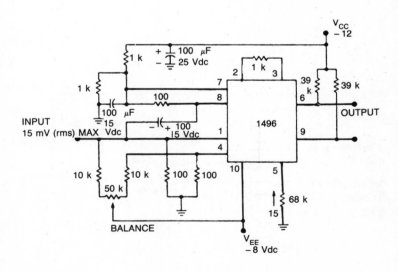

Low-frequency doubler

469

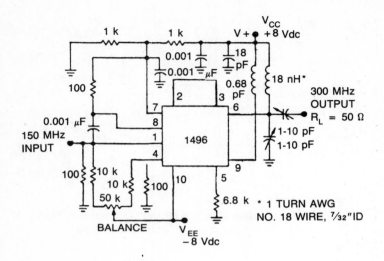

150 to 300 MHz doubler

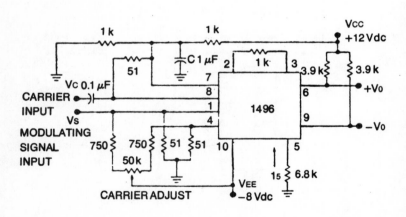

AM modulator circuit

470

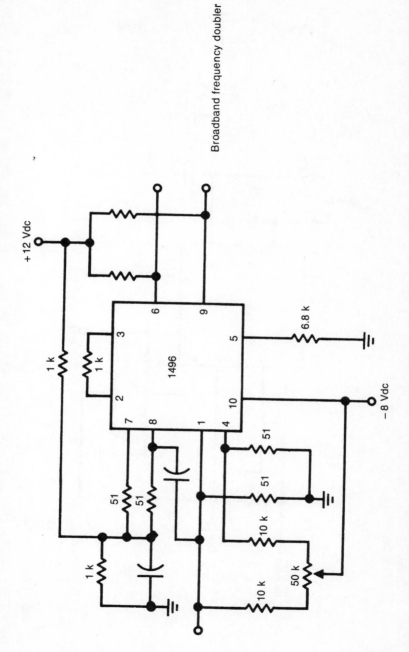

Broadband frequency doubler

471

SSB product detector

1590—WIDEBAND
INTEGRATED AMPLIFIER WITH A.G.C.

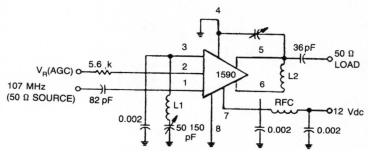

L1 = 24 TURNS, NO. 22 AWG WIRE ON A T12-44 MICRO
METAL TOROID CORE (– 124 pF)
L2 = 20 TURNS, NO. 22 AWG WIRE ON A T12-33 MICRO
METAL TOROID CORE (– 100 pF)

10.7-MHz amplifier gain $\simeq$ 55 dB, BW $\simeq$ 100 kHz

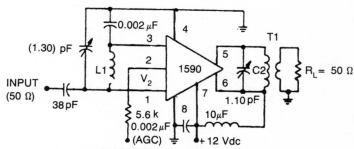

L1 = 12 TURNS #22 AWG WIRE ON A TOROID CORE.
(T37-6 MICRO METAL OR EQUIV)
T1: PRIMARY = 17 TURNS #20 AWG WIRE ON A TOROID CORE.
(T44-6 MICRO METAL OR EQUIV)
SECONDARY = 2 TURNS #20 AWG WIRE

30-MHz amplifier (Power Gain = 50 dB, BW $\approx$ 1.0 Mhz

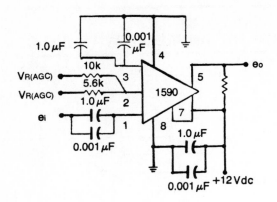

Video amplifier

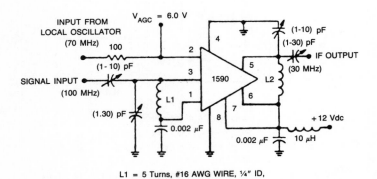

L1 = 5 Turns, #16 AWG WIRE, ¼" ID,
⅝ LONG
L2 = 16 TURNS, #20 AWG WIRE ON A TOROID
CORE. (T44-6 MICRO METAL OR EQUIV)

100-MHz mixer

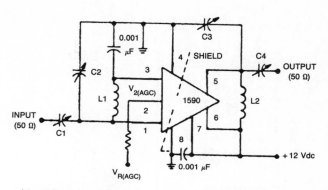

L1 = 7 Turns, #20 AWG Wire, 5/16" Dia.,
5/8" Long
L2 = 6 Turns, #14 AWG Wire, 9/16" Dia.,
+ E-F" Long

C1, C2, C3 = (1.30) pF
C4 = (1.10) pF

60-MHz power gain test circuit

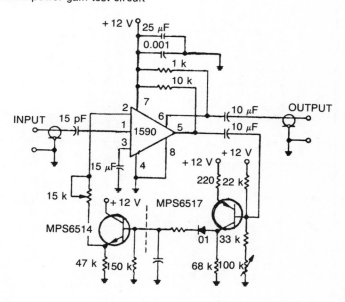

Speech compressor

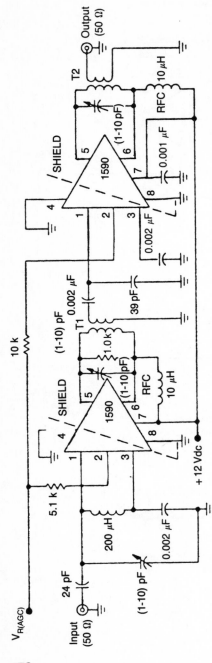

T1 PRIMARY WINDING = 15 TURNS, #22 AWG WIRE, ¼" IO AIR CORE
SECONDARY WINDING = 4 TURNS, #22 AWG WIRE,
COEFFICIENT OF COUPLING = 1.0

T2 PRIMARY WINDING = 10 TURNS, #22 AWG WIRE, ¼" IO AIR CORE
SECONDARY WINDING = 2 TURNS, #22 AWG WIRE,
COEFFICIENT OF COUPLING = 1.0

Two stage 60-MHz IF amplifier (Power gain ≈ 80 dB, BW ≈ 1.5 MHz)

1733—DIFFERENTIAL VIDEO WIDEBAND AMPLIFIER

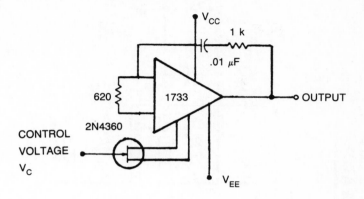

Voltage controlled oscillator

2111—INTEGRATED FM DETECTOR AND LIMITER

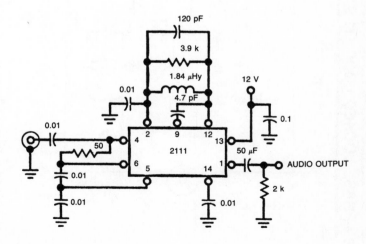

FM IF amplifier

477

1800—PHASE-LOCKED LOOP FM STEREO DEMODULATOR

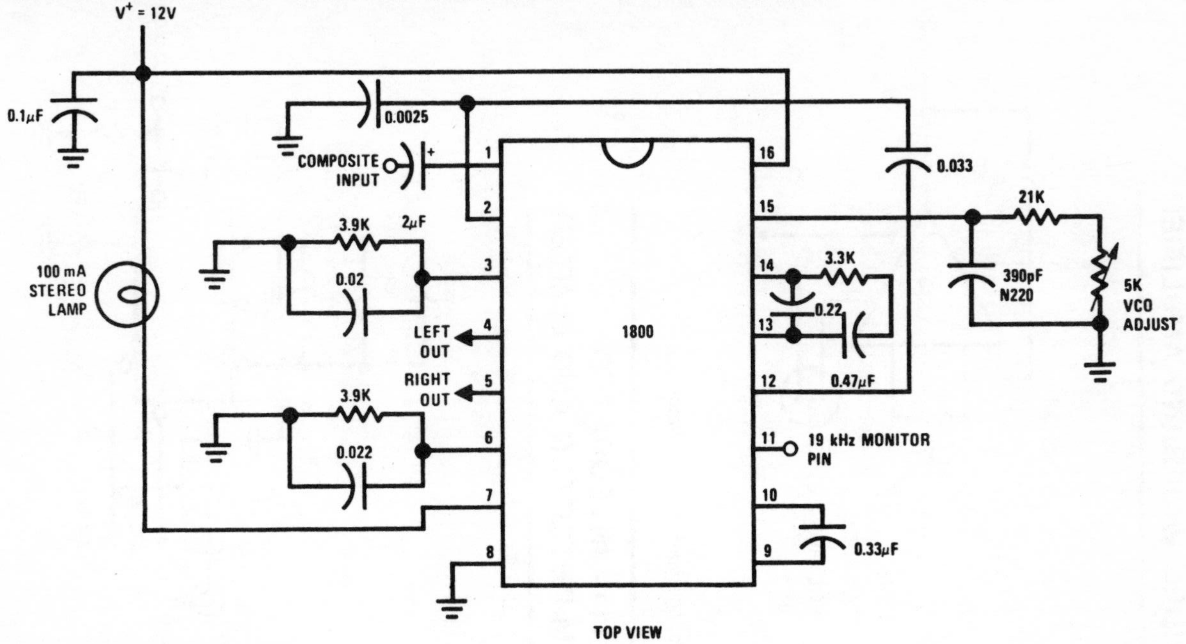

FM stereo demodulator with beacon

1889—INTEGRATED VIDEO MODULATOR

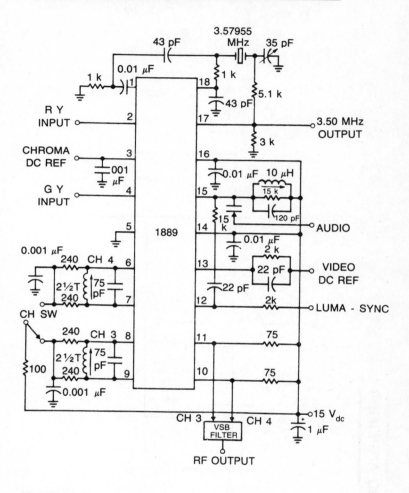

AC test circuit

3310—WIDEBAND AMPLIFIER

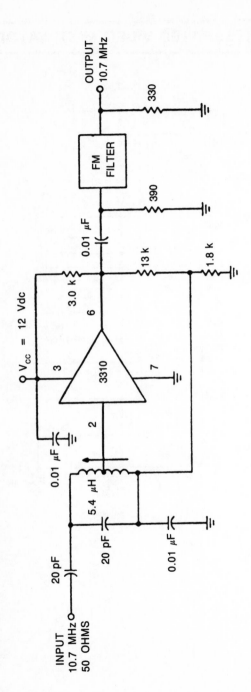

FM/IF amplifier

480

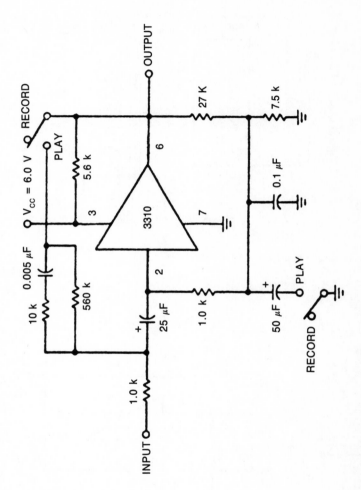

Record/Play preamplifier for cassette and portable tape recorders.

481

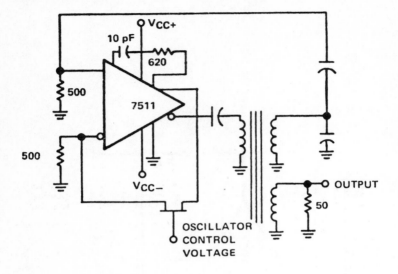

Basic transformer-feedback oscillator

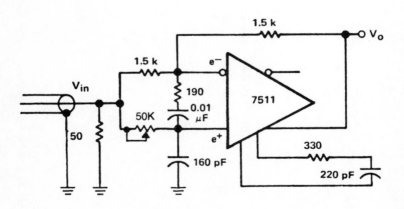

Variable-phase-shift circuit

482

Part 7
SPECIAL PURPOSE DEVICES

Sometimes you can't help but wonder if some circuits aren't designed more to amuse the design engineers than for a serious purpose! Whatever the reason, several interesting and novel ICs are now available. They can be used to produce many interesting and unusual projects.

MF10—DIGITAL MULTIMODE FILTER

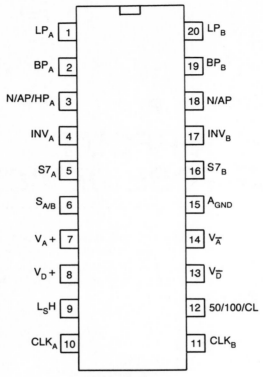

LP = LOW PASS
BP = BAND PASS
N = NOTCH (BAND REJECT)
H = HIGH PASS
CLK = CLOCK
AP = ALL PASS (PHASE SHIFT)

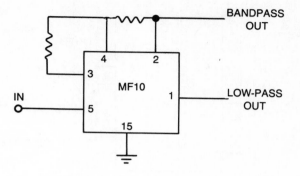

Low-pass/bandpass filter

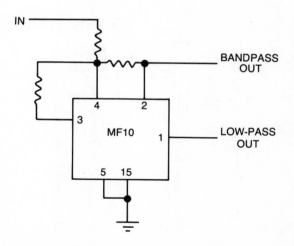

High Q low-pass/bandpass filter

485

567—TONE DECODER

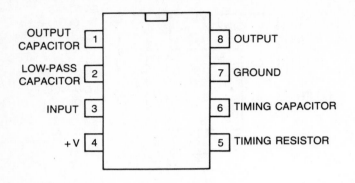

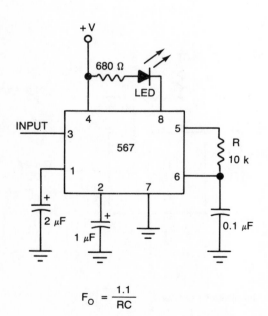

$$F_O = \frac{1.1}{RC}$$

Tone decoder

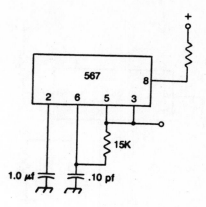

Oscillator with double frequency output

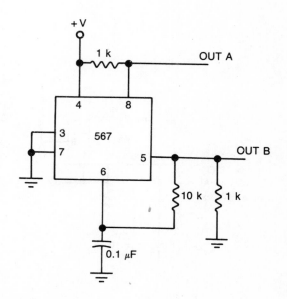

Two-phase oscillator

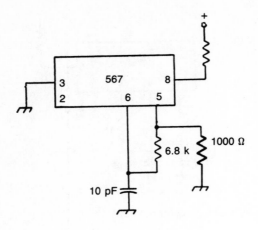

Oscillator with quadrature output

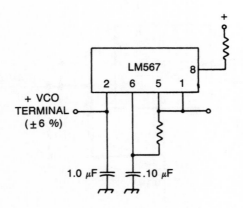

Precision oscillator drive 100 mA loads

488

570—COMPANDER

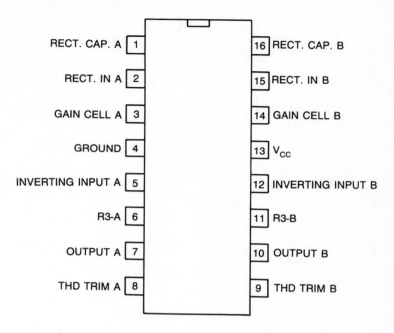

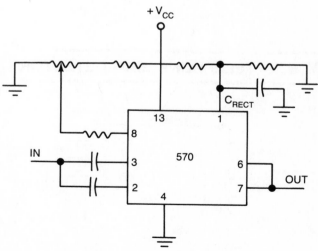

Compressor

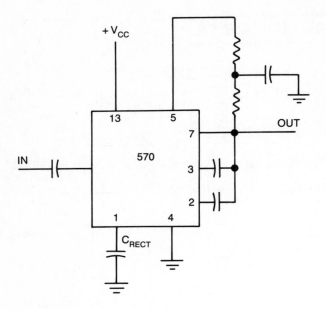

Expander

TCM1520—TELEPHONE RING DETECTOR

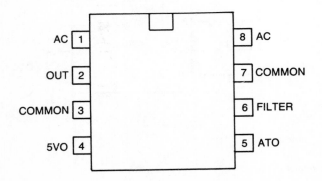

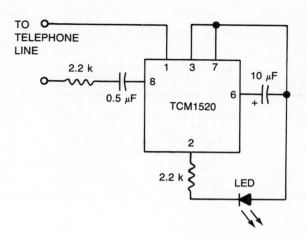

Visual ring detector

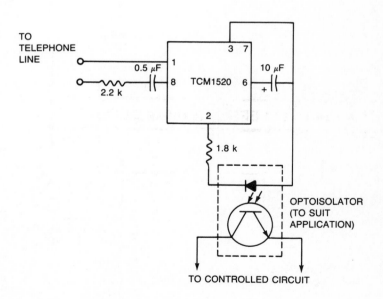

Ring activated optoisolator

TCM1532—RING DETECTOR/DRIVER

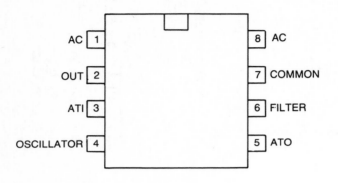

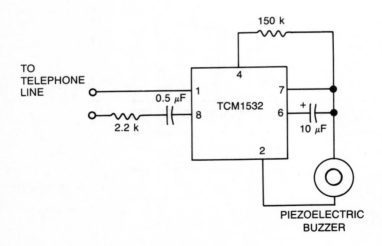

Remote ringer/buzzer

1913—TEMPERATURE TO FREQUENCY CONVERTER

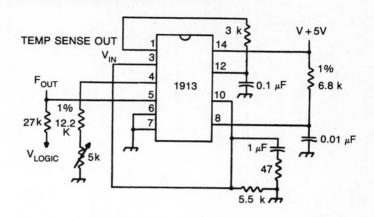

Fahrenheit scale temperature to frequency converter

2688—RANDOM NOISE GENERATOR

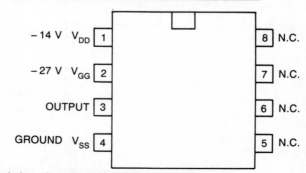

Fahrenheit scale temperature to frequency converter

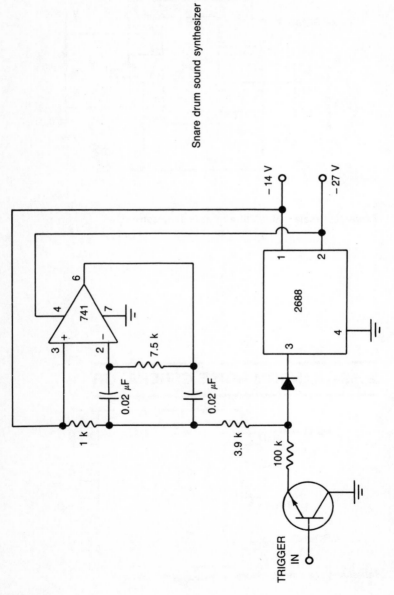

Snare drum sound synthesizer

Brush sound synthesizer

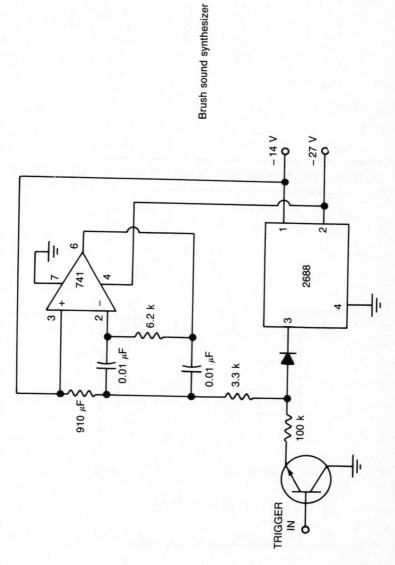

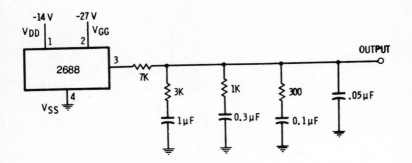

Pink noise generator

2907—FREQUENCY TO VOLTAGE CONVERTER

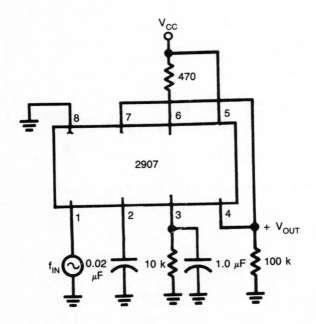

Zener regulated frequency to voltage converter

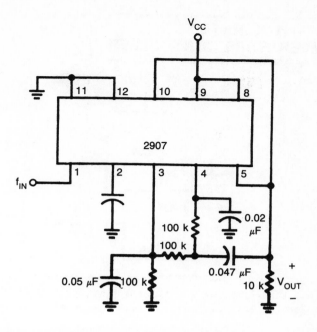

Frequency to voltage converter with 2 pole Butterworth filter to reduce ripple

3340—ELECTRONIC ATTENUATOR

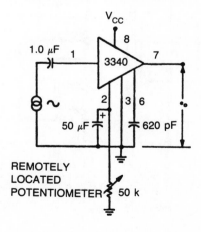

DC "remote" volume control

3911—INTEGRATED TEMPERATURE CONTROLLER

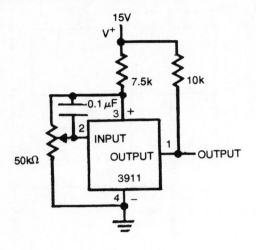

Basic temperature controller

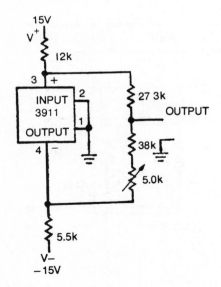

Ground referred centigrade thermometer

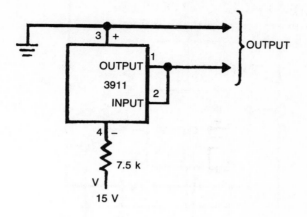

Basic thermometer for negative supply

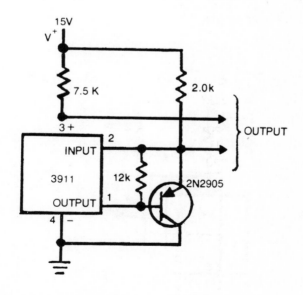

Increasing gain and output drive

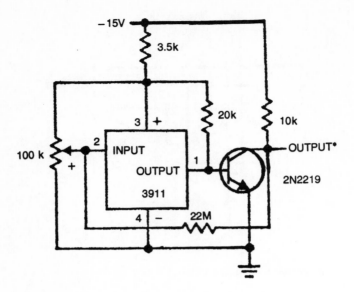

Temperature controller with hysteresis

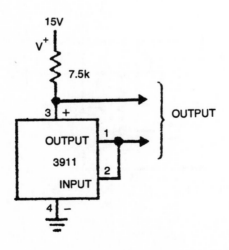

Basic thermometer for positive supply

500

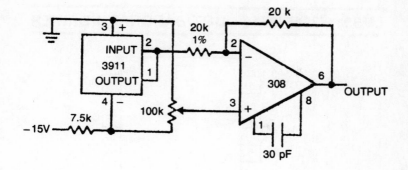

Ground referred centigrade thermometer

LM3914/3915/3916—DOT/BARGRAPH DRIVER

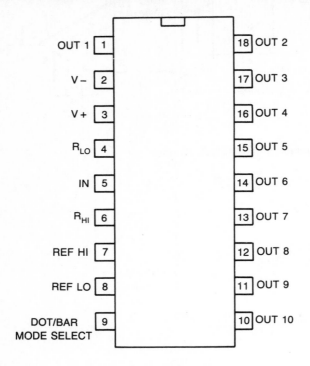

LM3914/ 3915/ 3916 DOT/BARGRAPH DISPLAY DRIVERS

These three chips are identical except for the values of the internal voltage divider resistances. The resistances in the LM3914 are matched for a linear response. The resistances in the LM3915 are set up for a logarithmic response. The LM3916 is designed for a semi-log scale and is intended for use in VU type applications.

While only the LM3914 is shown in the circuit diagrams, either the LM3915 or the LM3916 may be substituted with no other changes in the circuitry. These three ICs are fully pin-for-pin compatible.

Pin #9 is the mode select input. A HIGH signal (V+) on this pin causes the chip to operate in the bargraph mode. All of the output LEDS below the input voltage are lit. A LOW signal (ground) on pin #9 switches the IC into the dot mode. Only a single LED is lit to indicate the value of the input voltage. Lower valued LEDs are turned off.

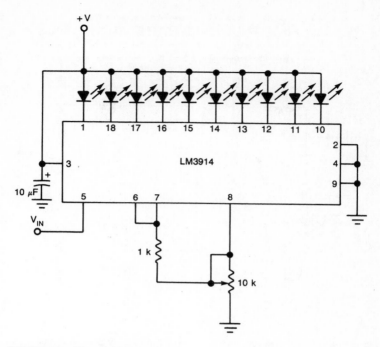

Dot display with variable range

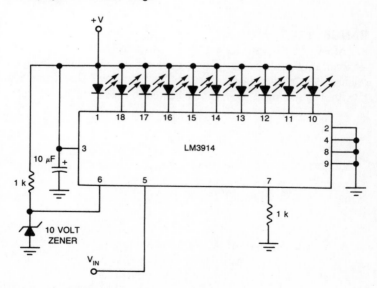

Precision 10 volt full-scale dot display

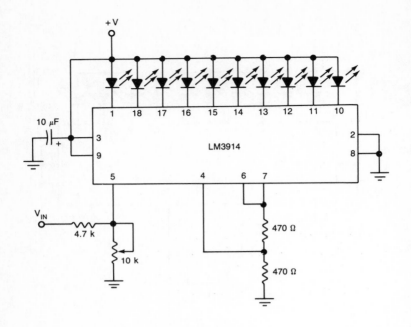

Range bargraph measures from X to Y volts, as set by potentiometer

RANGE BARGRAPH

Most LM3914 circuits measure a range from 0 volts up to some maximum full-scale value. This circuit has an adjustable minimum reading, set by the potentiometer. The circuit will read from X to Y volts. For example, it could be adjusted to read from 5 to 10 volts, or from 10 to 15 volts.

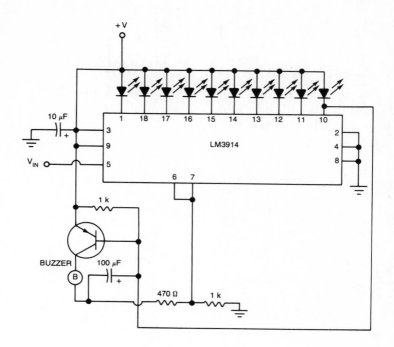

Bargraph with over-range alarm

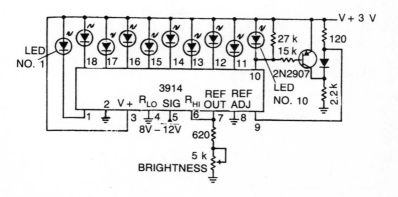

Indicator and alarm, full-scale changes display from dot to bar

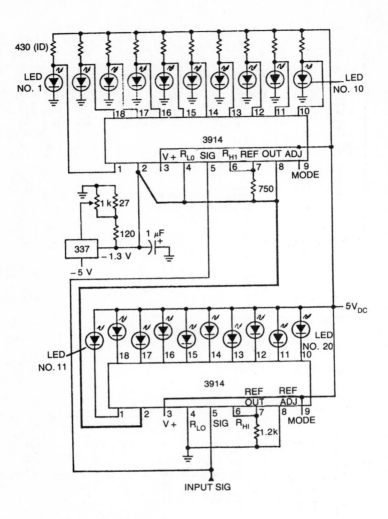

Zero-center meter, 20-segment

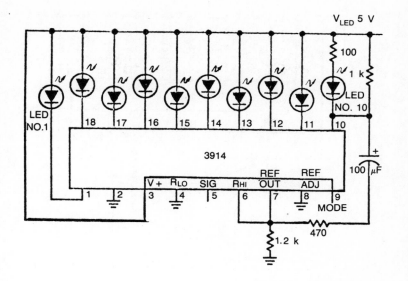

Bar display with alarm flasher

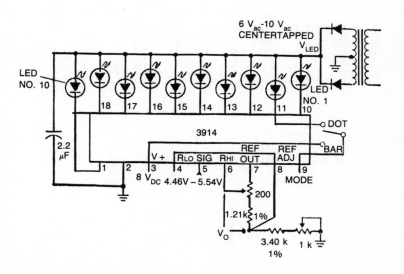

Expanded scale meter, dot or bar

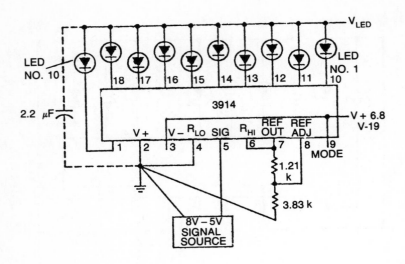

0 V to 5 V bargraph meter

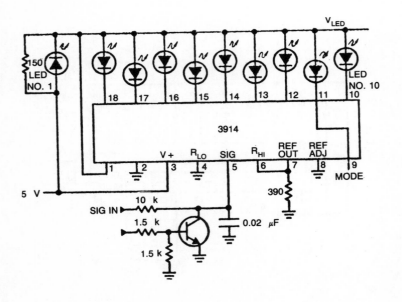

"Exclamation point" display

508

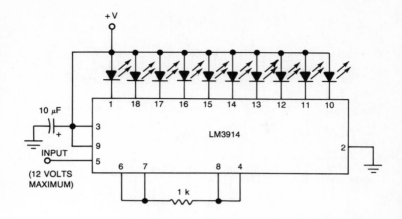

Bargraph display

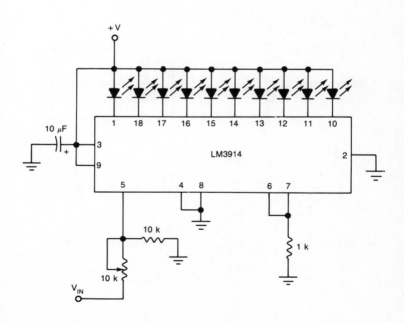

Bargraph display with adjustable full scale

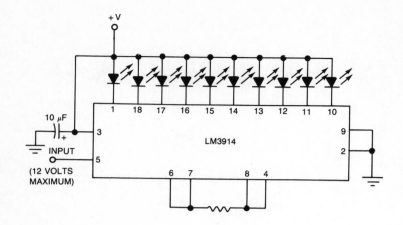

Dot graph display

5369—INTEGRATED
60 HZ TIMEBASE GENERATOR

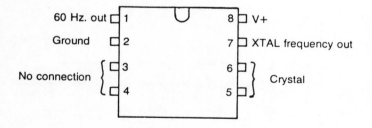

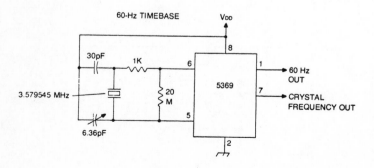

60-Hz timebase

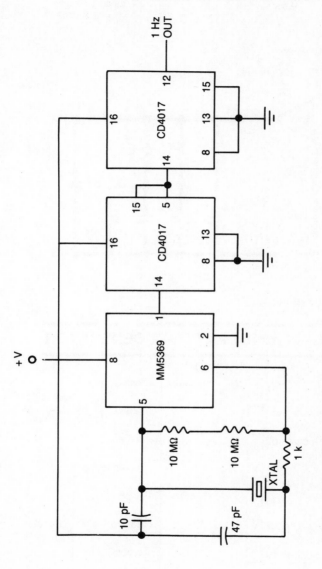

XTAL = 3.58 COLOR BURST CRYSTAL

One hertz timebase

511

5837—DIGITAL NOISE SOURCE INTEGRATED CIRCUIT

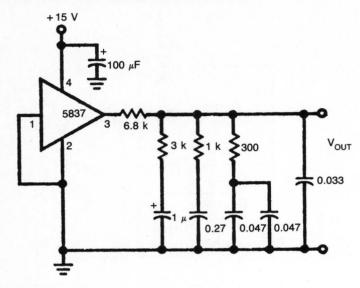

Pink noise generator

MM5871—RHYTHM PATTERN GENERATOR

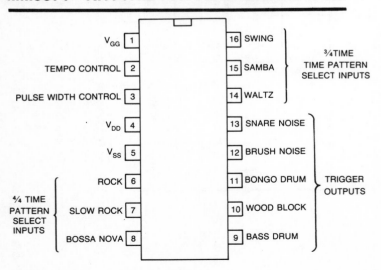

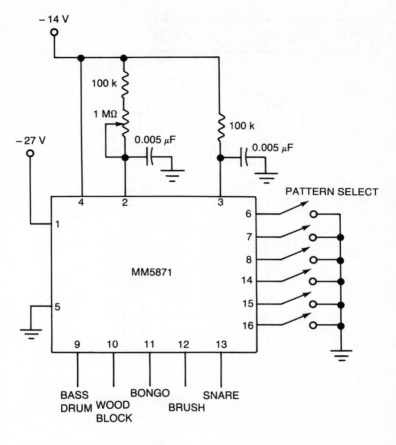

OUTPUTS TO DRUM
(OR OTHER) SYNTHESIZERS

Rhythm pattern generator

7910—MELODY GENERATOR

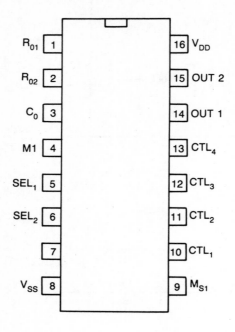

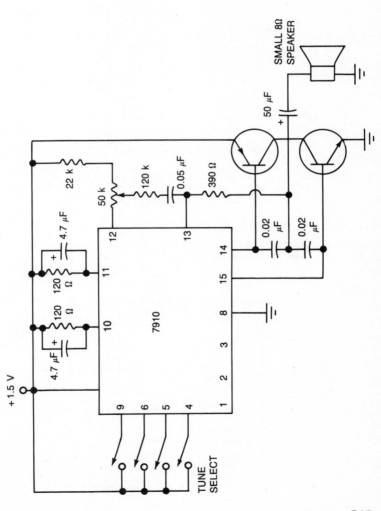

Melody generator

9400—FREQUENCY/VOLTAGE CONVERTER

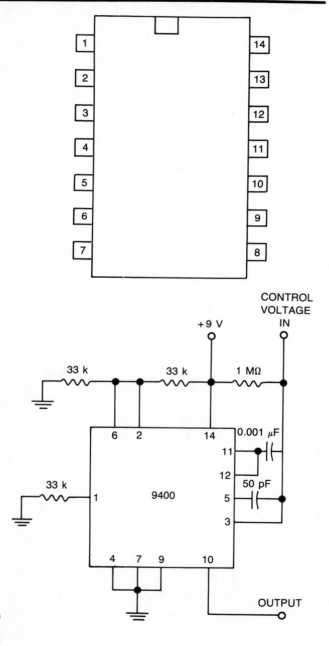

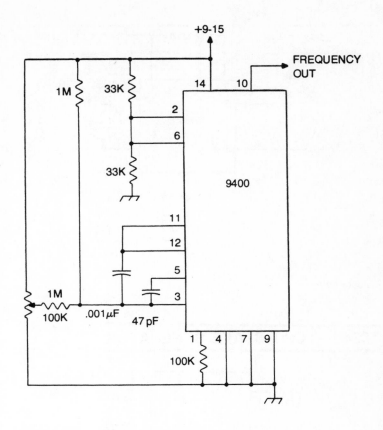

Basic voltage to frequency converter

MC34012—TELEPHONE RINGER

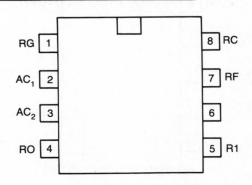

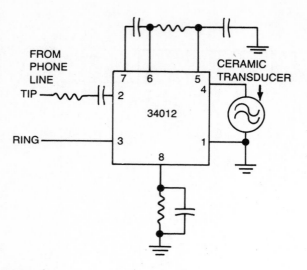

Remote ringer

50240—TOP OCTAVE GENERATOR

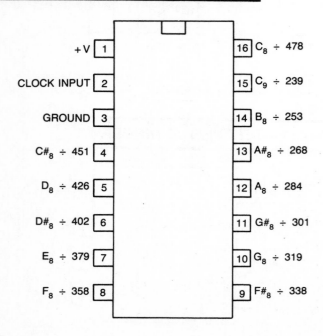

+V $\boxed{1}$	$\boxed{16}$ $C_8 \div 478$
CLOCK INPUT $\boxed{2}$	$\boxed{15}$ $C_9 \div 239$
GROUND $\boxed{3}$	$\boxed{14}$ $B_8 \div 253$
$C\#_8 \div 451$ $\boxed{4}$	$\boxed{13}$ $A\#_8 \div 268$
$D_8 \div 426$ $\boxed{5}$	$\boxed{12}$ $A_8 \div 284$
$D\#_8 \div 402$ $\boxed{6}$	$\boxed{11}$ $G\#_8 \div 301$
$E_8 \div 379$ $\boxed{7}$	$\boxed{10}$ $G_8 \div 319$
$F_8 \div 358$ $\boxed{8}$	$\boxed{9}$ $F\#_8 \div 338$

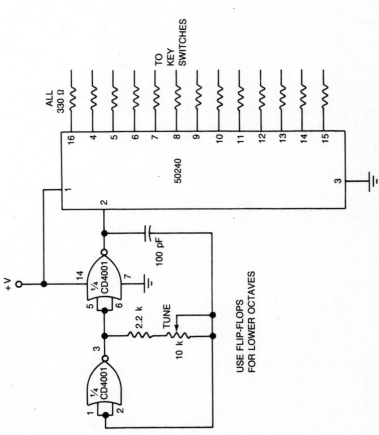

Full octave musical tone generator

519

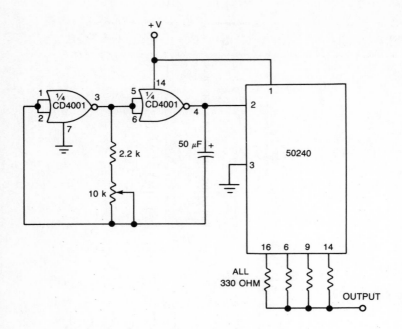

Random voltage generator

SN76477—COMPLEX SOUND GENERATOR

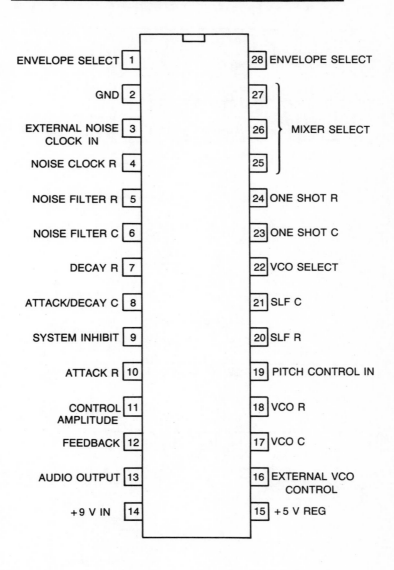

ENVELOPE SELECT — 1
GND — 2
EXTERNAL NOISE CLOCK IN — 3
NOISE CLOCK R — 4
NOISE FILTER R — 5
NOISE FILTER C — 6
DECAY R — 7
ATTACK/DECAY C — 8
SYSTEM INHIBIT — 9
ATTACK R — 10
CONTROL AMPLITUDE — 11
FEEDBACK — 12
AUDIO OUTPUT — 13
+9 V IN — 14

28 — ENVELOPE SELECT
27 — MIXER SELECT
26 — MIXER SELECT
25 —
24 — ONE SHOT R
23 — ONE SHOT C
22 — VCO SELECT
21 — SLF C
20 — SLF R
19 — PITCH CONTROL IN
18 — VCO R
17 — VCO C
16 — EXTERNAL VCO CONTROL
15 — +5 V REG

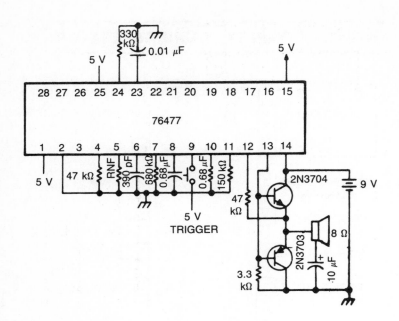

Gunshot/explosion

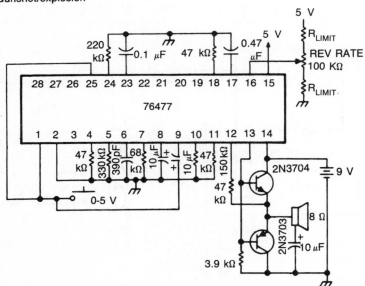

Race car motor/crash

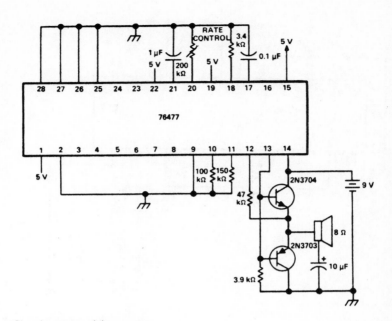

Siren/space war/phaser gun

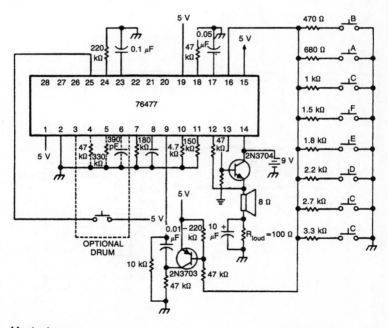

Musical organ

523

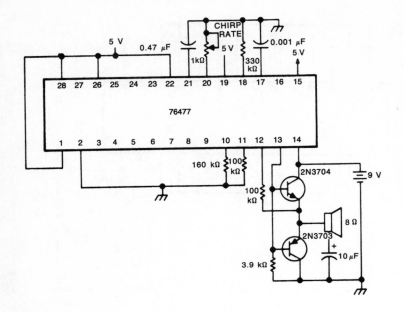

Bird chirp

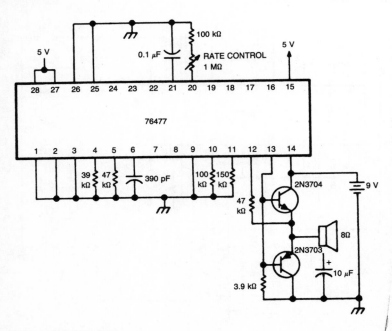

Steam train/prop plane

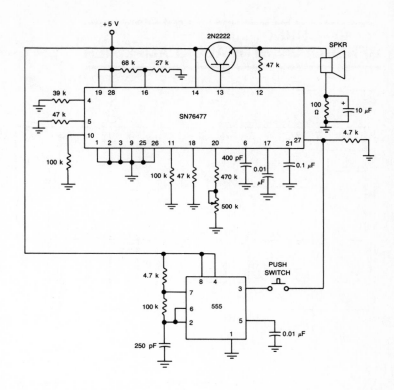

Steam train with whistle

76488—SOUND
GENERATOR WITH AUDIO AMPLIFIER

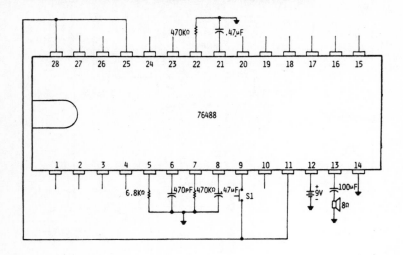

Gunshot

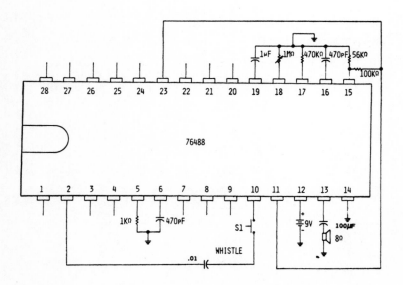

Toy steam engine and whistle

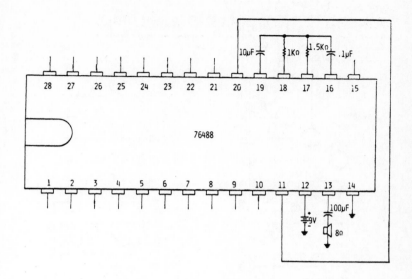

Phaser

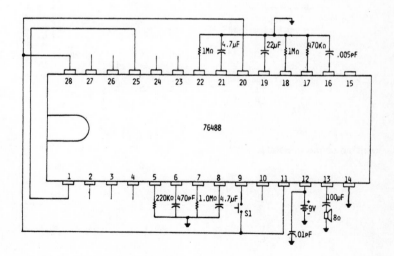

Bomb drop and explosion

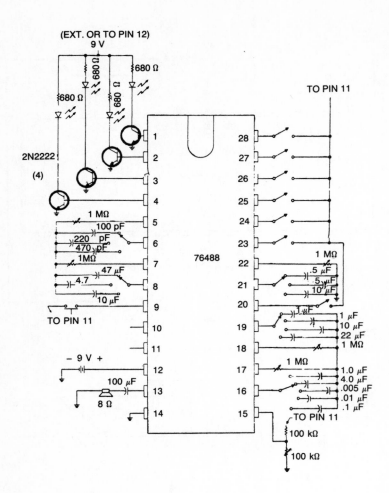

Multiple sound generator

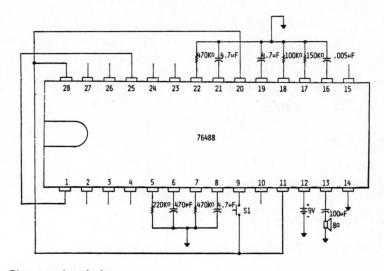

Phaser and explosion

SN94281—COMPLEX SOUND GENERATOR

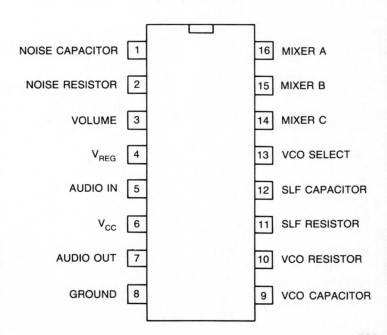

NOISE CAPACITOR — 1   16 — MIXER A

NOISE RESISTOR — 2   15 — MIXER B

VOLUME — 3   14 — MIXER C

V_{REG} — 4   13 — VCO SELECT

AUDIO IN — 5   12 — SLF CAPACITOR

V_{CC} — 6   11 — SLF RESISTOR

AUDIO OUT — 7   10 — VCO RESISTOR

GROUND — 8   9 — VCO CAPACITOR

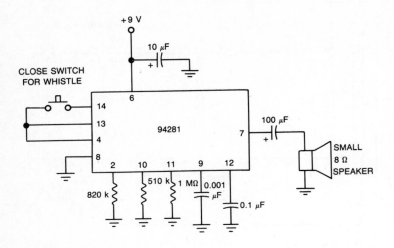

Close switch for whistle

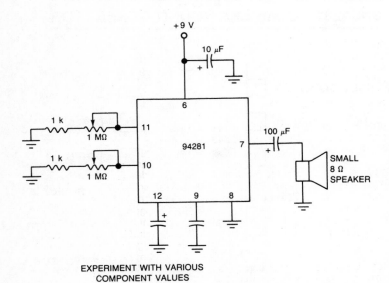

EXPERIMENT WITH VARIOUS
COMPONENT VALUES

Sound effect generator

Index